FUNCTIONAL ANALYSIS AND DIFFERENTIAL EQUATIONS IN ABSTRACT SPACES

SAMUEL ZAIDMAN

CHAPMAN & HALL/CRC
Monographs and Surveys in Pure and Applied Mathematics

CHAPMAN & HALL/CRC
Monographs and Surveys in
Pure and Applied Mathematics 100

FUNCTIONAL ANALYSIS AND DIFFERENTIAL EQUATIONS IN ABSTRACT SPACES

SAMUEL ZAIDMAN

CHAPMAN & HALL/CRC

Boca Raton London New York Washington, D.C.

Library of Congress Cataloging-in-Publication Data

Zaidman, Samuel, 1933-
Functional analysis and differential equations in abstract spaces / Samuel Zaidman.
p. cm. -- (Chapman & Hall/CRC monographs and surveys ; 100)
Includes bibliographical references and index.
ISBN 1-58488-011-2
1. Functional analysis. 2. Differential equations. 3. Hilbert space. 4. Banach spaces. I. Title. II. Series.
QA320.Z334 1999
515'.7—dc21 98-54425
CIP

Direct all inquiries to CRC Press LLC, 2000 N.W. Corporate Blvd., Boca Raton, Florida 33431.

International Standard Book Number1-58488-011-2
Library of Congress Card Number 98-54425
Printed in the United States of America 1 2 3 4 5 6 7 8 9 0
Printed on acid-free paper

Contents

Introduction

This is an elementary book, about (very) classical functional analysis, combined with some results on differential equations in Hilbert or Banach spaces, most of which are appearing for the first time in a monograph like this.

We cover in this book some part of *linear* functional analysis: the spaces where everything happens are Hilbert or Banach spaces; then, the operators acting therein are also linear, continuous or discontinuous.

The differential equations are of the form $Bu'(t) = Au(t)$ (so-called 'implicit' differential equations) in Chapter IV, and then in 'explicit' form (B = Identity). In Chapter IX it is a second-order equation: $u''(t) = Mu(t)$, written in 'weak' form in such a way that the second derivative actually disappears; here one discusses the Cauchy problem, essentially consisting of a search for solutions with given initial data. The next chapter deals with the Cauchy problem again. Now the equation is first-order: $u'(t) = Au(t)$, the space is Banach (more precisely a 'reflexive' Banach space) and the equation is in ultraweak form (somewhat akin to solutions in the theory of distributions) so that actually no derivative of the solution is required. In both Chapters IX and X the focus is on *uniqueness*: solutions with the same initial data must be equal.

Next we study, in Chapter XI, a different problem: the uniqueness of bounded solutions on the real line. Here the underlying space is again a Hilbert space; boundedness – in fact 'S^2-boundedness' – means that

$$\sup_{t\in\mathbb{R}} \int_t^{t+1} \|u(\sigma)\|^2 \, d\sigma < \infty ;$$

the solution is again in the ultraweak tense. The result established in this section, a mild extension of a previous theorem of the present author [15], appears here for the first time in the mathematical literature.

INTRODUCTION

Chapter XII deals with some properties of almost-periodic solutions. The equation is now of the 'non-homogeneous' kind: $v'(t) = Av(t) + Bv(t) + g(t)$, for continuously differentiable solutions $v(\cdot)$ in a Banach space; the operator A is now the 'infinitesimal generator' of a group of operators $G(t)$ (concepts defined in the course of the book!). The function $g(\cdot)$ is almost-periodic, as is the group $G(\cdot)$; the operator B is a compact operator. What we prove is a logical implication: solutions $v(\cdot)$ which are almost-periodic in the weak sense are almost-periodic in the strong (usual) sense as well. There is also another chapter dealing with almost-periodic solutions, namely, Chapter XV. Here the equation is $u'(t) - Au(t) = f(t)$, on the real line; the operator A is a generator of a C_0-semigroup of operators (a concept studied earlier in the book). One assumes that $f(\cdot)$ and $u(\cdot)$ are almost-periodic from $\mathbb{R}$ to the Banach space X and that $u(t)$, not necessarily differentiable, is a 'mild' solution, as defined in past work of the author (it admits a certain representation formula in every half-line $[a,+\infty)$, involving the semigroup whose generator is A). It is proved that, $\forall \lambda \in \mathbb{R}$, the mean value

$$a(\lambda, u) = \lim_{r \to \infty} \frac{1}{r} \int_0^r \mathrm{e}^{-\mathrm{i}\lambda t} u(t)\, \mathrm{d}t \,,$$

belongs to the definition domain $\mathcal{D}(A)$ and that $(\mathrm{i}\lambda - A)a(\lambda, u) = a(\lambda, f)$ holds true.

Returning to the Cauchy problem we have Chapter XIII, where an ultraweak situation in a reflexive B space is examined. One assumes pointwise continuous dependence on the initial data, which entails the existence of an operator semigroup $U(t)$ transporting the initial data in the solution at time t. The equation is again $u'(t) - Au(t) = \mathbf{0}$ on $[0, \infty)$ and it is shown that there exists a real number ω, such that the classical equality

$$\int_0^\infty \mathrm{e}^{-\lambda t} U(t)x \, \mathrm{d}t = (\lambda I - A)^{-1} x \,, \quad \forall x \in \mathcal{D}(A) \,, \quad \forall \lambda \in \mathbb{C}, \ \operatorname{Re} \lambda > \omega$$

is now valid in the 'ultraweak Cauchy problem' situation.

Finally, in Chapter XIV, we point out a result about ultraweak differential inequalities in Hilbert space. The strong form of this inequality is $\|u'(t) - Au(t)\| \le \phi(t)\|u(t)\|$, $0 \le t < \infty$; the corresponding ultraweak version is defined in Chapter XIV. Under a uniform boundedness condition for the resolvent operator $(\lambda I - A)^{-1}$ it is proved that solutions which decrease faster than any exponential are identically null on $[0, \infty)$. The main step in the proof concerns the passage from the ultraweak case to the regular (C^1) situation which is effected by the mollification method.

We conclude this introduction with a remark about the quite well-known section on general functional analysis. This appears in several separate chapters but also as essential components in some chapters on differential equations. Also, almost everything is classical; however, here and there, the attentive reader will note things difficult to find elsewhere (take for instance the result about the unique extension of linear continuous functionals from a linear subspace of a Hilbert space to the whole space (Theorem 5.1, Chapter I).

We conclude the book with an index of notation, a subject index, and a list of bibliographical references. They should be useful to the careful reader.

Chapter I

Hilbert space, definitions, and first results

Introduction

Hilbert spaces are a natural generalization of the euclidean space $\mathbb{R}^n$. They usually have infinite dimension and appear as the best setting for various problems in mathematics and in mathematical physics.

Our basic object is a vector (= linear) space V over the real field $\mathbb{R}$ or the complex field $\mathbb{C}$, endowed with a 'scalar' (or 'inner') product between its elements – to any couple of vectors $x, y \in V$, one associates a number (x, y) in such a way that the following properties are satisfied:

(i) $(x,x) \geq 0$ and $(x,x) = 0$ iff $x = \mathbf{0}$ (the 'zero' element in V).

(ii) $(x, y+z) = (x,y) + (x,z), \ \forall x,y,z \in V.$ (0.1)

(iii) $(x, \lambda y) = \overline{\lambda}(x,y), \ \forall x, y \in V$ and $\forall \lambda \in K$ (where K means $\mathbb{R}$ or $\mathbb{C}$).

(iv) $(x,y) = \overline{(y,x)}, \ \forall x,y \in V.$

Let us note that from (i)-(iv) we can easily derive:

(1) $(x, \alpha y + \beta z) = (x, \alpha y) + (x, \beta z) = \overline{\alpha}(x,y) + \overline{\beta}(x,z), \ \forall x,y,z \in V, \ \alpha, \beta \in K.$

(2) $(\lambda x, y) = \overline{(y, \lambda x)} = \overline{\overline{\lambda}(y,x)} = \lambda\overline{(y,x)} = \lambda(x,y), \ \forall x,y \in V, \ \lambda \in K.$ (0.2)

(3) $(x+y, z) = \overline{(z, x+y)} = \overline{(z,x) + (z,y)} = \overline{(z,x)} + \overline{(z,y)} = (x,z) + (y,z), \ \forall x,y,z \in V.$

Examples

(1) Form the cartesian product $\mathbb{C}^n = \mathbb{C} \times \mathbb{C} \times \ldots \times \mathbb{C}$ (n-times). Then, if $x = \langle x_1, x_2, \ldots x_n \rangle$, $y = \langle y_1, y_2, \ldots, y_n \rangle$ are elements in this space we define

$$(x,y) = \sum_{i=1}^{n} x_i \overline{y}_i \ . \tag{0.3}$$

(2) Consider $C[a,b]$, the linear space of all continuous complex-valued functions on $[a,b]$. Then, for $f,g \in C[a,b]$, we define

$$(f,g) = \int_a^b f(x)\overline{g}(x)\,\mathrm{d}x \ . \tag{0.4}$$

1. Orthogonality

We say that elements x,y in V are orthogonal if their inner product (x,y) is zero. Thus we write:

$$x \perp y \quad \text{iff} \quad (x,y) = 0 \ . \tag{1.1}$$

Likewise one defines orthonormal sets in V by the following property:

$$\begin{aligned} A \subset V \text{ is orthonormal iff } (x,y) &= 0 \quad \forall x,y \in A,\ x \neq y \\ \text{and } (x,x) &= 1 \quad \forall x \in A \ . \end{aligned} \tag{1.2}$$

The *norm* $\|x\|$ of an element in V is defined by the formula

$$(x,x)^{\frac{1}{2}} = \|x\| \ . \tag{1.3}$$

An important result, known as 'Bessel's inequality' can be stated as follows:

Proposition 1.1 *Let* $\{x_n\}_{n=1}^N$ *be a (finite) orthonormal set in V. Then,* $\forall x \in V$, *the following inequality*

$$\sum_{n=1}^N \left|(x,x_n)\right|^2 \le \|x\|^2 \tag{1.4}$$

holds true.

Proof Let us start with the (obvious) decomposition of the element x:

$$x = \sum_{n=1}^N (x,x_n)x_n + \left(x - \sum_{n=1}^N (x,x_n)x_n \right) = y + (x-y) \ . \tag{1.5}$$

Next note that the elements y and $x - y = z$ are orthogonal:

$$\left(\sum_{n=1}^{N}(x,x_n)x_n,\ x-\sum_{n=1}^{N}(x,x_n)x_n\right)$$

$$=\left(\sum_{n=1}^{N}(x,x_n)x_n, x\right)-\left(\sum_{n=1}^{N}(x,x_n)x_n,\ \sum_{n=1}^{N}(x,x_n)x_n\right) \quad \text{(using (0.1) and (0.2))}$$

$$=\sum_{n=1}^{N}(x,x_n)(x_n,x)-\sum_{i,j=1}^{N}\left((x,x_i)x_i,\ (x,x_j)x_j\right)$$

$$=\sum_{n=1}^{N}\left|(x,x_n)\right|^2-\sum_{i=1}^{N}(x,x_i)\cdot\overline{(x,x_i)}=0\ .$$

Therefore we get that $\|z\|^2+\|y\|^2=\|y+z\|^2$, hence $\|y\|^2\leq\|y+z\|^2=\|x\|^2$. Actually

$$\|y\|^2=\left(\sum_{n=1}^{N}(x,x_n)x_n,\ \sum_{n=1}^{N}(x,x_n)x_n\right)$$

$$=\sum_{n=1}^{N}(x,x_n)\cdot\overline{(x,x_n)}$$

$$=\sum_{n=1}^{N}\left|(x,x_n)\right|^2$$

and we obtain (1.4). ■

As a corollary we derive an equally fundamental result known as the 'Cauchy-Schwarz inequality':

Proposition 1.2 *For any pair of elements x, y in V, the following holds*:

$$|(x,y)|\leq\|x\|\cdot\|y\|\ . \tag{1.6}$$

Proof If $y=\mathbf{0}$, we have $(x,y)=(x,y)+(x,y)$, hence $(x,y)=0$. Also $\|y\|=0$ and (1.6) is (trivially) true.

Otherwise we can consider the single-element set $\left\{ y/\|y\| \right\}$ which appears as an orthonormal set. Applying (1.4) we obtain

$$\left|\left(x, \frac{y}{\|y\|}\right)\right|^2 \leq \|x\|^2 \tag{1.7}$$

which becomes

$$\frac{|(x,y)|^2}{\|y\|^2} \leq \|x\|^2 , \qquad |(x,y)| \leq \|x\|\,\|y\|$$

that is (1.6). ■

Another useful result appears as

Proposition 1.3 *(The parallelogram law). For any pair of elements* $x, y \in V$, *the following equality holds:*

$$\|x+y\|^2 + \|x-y\|^2 = 2\left(\|x\|^2 + \|y\|^2\right) . \tag{1.8}$$

Proof We have in fact

$$\begin{aligned}\|(x+y)\|^2 + \|x-y\|^2 &= (x+y, x+y) + (x-y,\ x-y) \\ &= 2\|x\|^2 + 2\|y\|^2 + (x,y) + (y,x) - (y,x) - (x,y) \\ &= 2\left(\|x\|^2 + \|y^2\|\right) .\end{aligned}$$ ■

Note. A linear space over $\mathbb{C}$ or $\mathbb{R}$ with a scalar product is called a pre-Hilbert space.

Normed linear spaces

These are again linear spaces V over a field K (which is $\mathbb{R}$ or $\mathbb{C}$) endowed with a function ('norm'), $V \to \mathbb{R}$, denoted by: $x \in V \to \|x\| \in \mathbb{R}$, in such a way that

$$\begin{aligned}&\|x\| \geq 0 \quad \forall x \in V \quad \text{and} \quad \|x\| = 0 \quad \text{iff} \quad x = \mathbf{0} \\ &\|\alpha x\| = |\alpha|\,\|x\| \quad \forall \alpha \in K,\ x \in V \\ &\|x+y\| \leq \|x\| + \|y\|, \quad \forall x, y \in V .\end{aligned} \tag{1.9}$$

If we now refer to definition (1.3), we (have to) prove the following:

Proposition 1.4 *Any pre-Hilbert space V becomes a linear normed space under definition* (1.3): $\|x\| = (x,x)^{\frac{1}{2}}$, $\forall x \in V$.

Proof First we note that, $\forall \alpha \in K$, $x \in V$, we have

$$\|\alpha x\|^2 = (\alpha x, \alpha x) = \alpha \cdot \overline{\alpha} \|x\|^2 = |\alpha|^2 \|x\|^2 \ ;$$

furthermore, $\forall x, y \in V$

$$\|x+y\|^2 = (x+y, x+y) = \|x\|^2 + \|y\|^2 + 2\,\mathrm{Re}(x,y) \leq \|x\|^2 + \|y\|^2 + 2\|x\|\,\|y\|$$

(by (1.6)); thus

$$\|x+y\|^2 \leq (\|x\| + \|y\|)^2 \ . \qquad \blacksquare$$

Remark Any linear normed space V is also a metric (linear) space, where the distance between elements $x, y \in V$ is defined by

$$d(x,y) = \|x-y\| \ . \tag{1.10}$$

Therefore the concepts of a convergent sequence, a Cauchy sequence, completeness ('any Cauchy sequence is convergent = has a limit'), and density of subsets, which are known in metric space, have the corresponding meaning in linear normed spaces, as well as in pre-Hilbert spaces. In particular, we shall give the following:

Definition *Any complete pre-Hilbert space is a Hilbert space. Any complete linear-normed space is a Banach space.*

2. Geometry in Hilbert spaces

Let $\mathcal{H}$ be a Hilbert space and then $\mathcal{M}$ be a closed (linear) subspace of $\mathcal{H}$ (that is, if $(x_n)_1^\infty$ is a sequence in $\mathcal{M}$ and $x_n \to x_0 \in \mathcal{H}$, then $x_0 \in \mathcal{M}$).

With the induced scalar product $\mathcal{M}$ is a pre-Hilbert subspace of $\mathcal{H}$ which is also complete (if $(x_n)_1^\infty$ is a Cauchy sequence in $\mathcal{M}$ it converges to some $\bar{x} \in \mathcal{H}$ and then, by closedness, $\bar{x} \in \mathcal{M}$, too).

Let us now define in $\mathcal{H}$ the subset $\mathcal{M}^\perp$ ('the orthogonal complement to $\mathcal{M}$') which is given by the relation

$$\mathcal{M}^\perp = \{h \in \mathcal{H},\ h \perp \mathcal{M}\} = \{h \in \mathcal{H},\ (h,m) = 0\ \forall m \in \mathcal{M}\}\ . \tag{2.1}$$

Now, the subset $\mathcal{M}^\perp$ is again a linear subspace of $\mathcal{H}$ - this is readily seen. Furthermore, $\mathcal{M}^\perp$ is also a *closed* linear subspace: in fact, if $h_n \in \mathcal{M}^\perp$ and $h_n \to h_0$, it follows that $(h_n, m) = 0\ \forall m \in \mathcal{M}$ (and $\forall n \in \mathbb{N}$) and then $(h_0, m) = 0\ \forall m \in \mathcal{M}$: this last property holds because

$$|(h_n,m) - (h_0,m)| = |(h_0,m)| \le \|h_n - h_0\|\,\|m\| \to 0 \quad \text{as } n \to \infty\ .$$

Thus, we now have two Hilbert subspaces of $\mathcal{H}$, $\mathcal{M}$, and $\mathcal{M}^\perp$. Note that $\mathcal{M} \cap \mathcal{M}^\perp = \{\mathbf{0}\}$ (if $h \in \mathcal{M} \cap \mathcal{M}^\perp$ then $(h,h) = 0$!).

A very fundamental fact of Hilbert space theory is the *projection theorem*:

Theorem 2.1 *Let $\mathcal{M}$ be a closed linear subspace of a Hilbert space $\mathcal{H}$. Any $h \in \mathcal{H}$ can be written in the form $h = k + l$, where $k \in \mathcal{M}$, $l \in \mathcal{M}^\perp$. The elements k, l are uniquely determined by h.*

(We say that $\mathcal{H}$ is the orthogonal direct sum of $\mathcal{M}$ and $\mathcal{M}^\perp$.)

Proof We first present:

Lemma 2.1 *Let H be a Hilbert space and $K \subset H$ be a closed convex set. Denote: $d = \inf\{\|x\|,\ x \in K\}$. Then, there exists a single element $x_0 \in K$ such that $\|x_0\| = d$.*

Proof of Lemma Note first that, $\forall n \in \mathbb{N}$, $\exists x_n \in K$, such that

$$d \le \|x_n\| < d + \frac{1}{n}\ ; \qquad \text{thus,} \qquad \lim \|x_n\| = d\ . \tag{2.2}$$

Next we shall see that the sequence (x_n) is a Cauchy sequence: using the parallelogram law we get the relation

$$\|x_n - x_m\|^2 = 2\left(\|x_n\|^2 + \|x_m\|^2\right) - \|x_n + x_m\|^2 \tag{2.3}$$

which is also expressed as

$$\left\|\frac{x_n - x_m}{2}\right\|^2 = \frac{1}{2}\left(\|x_n\|^2 + \|x_m\|^2\right) - \left\|\frac{x_n + x_m}{2}\right\|^2 . \tag{2.4}$$

From the convexity of the set K we get: $\frac{1}{2}(x_n + x_m) \in K$, hence, $\left\|\frac{1}{2}(x_n + x_m)\right\| \geq d$. Accordingly, from (2.4), we derive the inequality:

$$\left\|\frac{x_n - x_m}{2}\right\|^2 \leq \frac{1}{2}\left(\|x_n\|^2 + \|x_m\|^2\right) - d^2 . \tag{2.5}$$

From (2.2) we get $\|x_n\|^2 \to d^2$, $\|x_m\|^2 \to d^2$; hence, $\forall \varepsilon > 0$, we have $\|x_n\|^2 < d^2 + \varepsilon^2$, $\|x_m\|^2 < d^2 + \varepsilon^2$, if $n \geq \overline{n}(\varepsilon)$, $m \geq \overline{n}(\varepsilon)$. Thus (2.5) now gives:

$$\|x_n - x_m\|^2 < 2\left(2d^2 + 2\varepsilon^2\right) - 4d^2 = 4\varepsilon^2 , \qquad \text{for } n, m \geq \overline{n}(\varepsilon) , \tag{2.6}$$

which means that $(x_n)_1^\infty$ is a Cauchy sequence. Let $x_0 = \lim x_n$. Then $x_0 \in K$ and $\|x_0\| = \lim\|x_n\| = d$.

Finally, suppose that $x_1 \in K$ and $\|x_1\| = d$. Again, $\frac{1}{2}(x_0 + x_1) \in K$ and also

$$0 \leq \left\|\frac{x_1 - x_0}{2}\right\|^2 = \frac{1}{2}\left(\|x_0\|^2 + \|x_1\|^2\right) - \left\|\frac{x_0 + x_1}{2}\right\|^2 = d^2 - \left\|\frac{x_0 + x_1}{2}\right\|^2 \leq 0 . \tag{2.7}$$

Thus, $\|x_1 - x_0\| = 0$, $x_1 = x_0$. ■

The next step towards Theorem 2.1 is the following:

Corollary 2.1 *Let $\mathcal{H}$ be a Hilbert space, $\mathfrak{M}$ a closed (linear) subspace of $\mathcal{H}$ and $x \in \mathcal{H}$. Then, there exists in $\mathfrak{M}$ a unique element z such that*

$$\|z - x\| \leq \|y - x\| \quad \forall y \in \mathfrak{M} . \tag{2.8}$$

In fact, let $K = x - \mathcal{M} = \{x - m,\ m \in \mathcal{M}\}$. It is easy to see that K is a closed convex set in $\mathcal{H}$. Hence, $\exists! w \in K$, such that

$$\|w\| = \inf_{m \in \mathcal{M}} \|x - m\| \ . \tag{2.9}$$

Now, $w \in K$ means that $w = x - z$, for some $z \in \mathcal{M}$. Thus

$$\|x - z\| \le \|x - y\| , \quad \forall y \in \mathcal{M} \ .$$ ■

We are now ready to present the final step towards Theorem 2.1. One first establishes the uniqueness of the decomposition: therefore, let us assume that

$$h = k_1 + l_1 = k_2 + l_2 , \quad \text{where } \ k_i \in \mathcal{M},\ l_i \in \mathcal{M}^{\perp},\ \forall i = 1,2 \ . \tag{2.10}$$

It will follow that

$$k_1 - k_2 = l_2 - l_1 \quad \text{and} \quad k_1 - k_2 \perp l_2 - l_1 \ .$$

This entails $k_1 = k_2$, $l_1 = l_2$.

Next, we prove the existence of the decomposition. For $h \in \mathcal{H}$ we consider (according to Corollary 2.1) the unique element $z \in \mathcal{M}$, such that

$$\|z - h\| \le \|y - h\| \quad \forall y \in \mathcal{M} \ . \tag{2.11}$$

Write then $h = z + (h - z)$. We show that $h - z \in \mathcal{M}^{\perp}$. Denote $h - z = w$. From (2.11) we derive

$$d = \|z - h\| \le \|h - (z + tu)\| , \quad \forall t \in \mathbb{R},\ \forall u \in \mathcal{M},$$

that is, also,

$$d^2 \le \|w - tu\|^2 = \|w\|^2 - 2t\,\mathrm{Re}(w,u) + t^2\|u\|^2 = d^2 - 2t\,\mathrm{Re}(w,u) + t^2\|u\|^2 ,$$

which gives, accordingly, the inequality

$$0 \le -2t\,\mathrm{Re}(w,u) + t^2\|u\|^2\,, \qquad \forall t \in \mathbb{R}$$

or

$$t\left(t\|u\|^2 - 2\,\mathrm{Re}(w,u)\right) \ge 0\,, \qquad \forall t \in \mathbb{R}\,. \tag{2.12}$$

Now, if $\mathrm{Re}(w,u) \neq 0$, the inequality (2.12) fails for $0 < t < \dfrac{2\,\mathrm{Re}(w,u)}{\|u\|^2}$ or for $\dfrac{2\,\mathrm{Re}(w,u)}{\|u\|^2} < t < 0$. Thus, $\mathrm{Re}(w,u) = 0$, $\forall u \in \mathcal{M}$, and in Hilbert spaces over $\mathbb{R}$ we are done.

Otherwise, in Hilbert spaces over $\mathbb{C}$, repeating the previous reasoning with t replaced by $i\tau$, $\tau \in \mathbb{R}$, we obtain $\mathrm{Im}(w,u) = 0\ \forall u \in \mathcal{M}$. In any case we get that $(w,u) = 0\ \forall u \in \mathcal{M}$, $w \perp \mathcal{M}$. ■

Our final result in this section is:

Theorem 2.2 *Let $\mathcal{M} \subset \mathcal{H}$ be a linear (not necessarily closed) subspace of the Hilbert space $\mathcal{H}$, and $\overline{\mathcal{M}}$ be the closure of $\mathcal{M}$ (in the metric space $\mathcal{H}$). Then the following holds:*

$$\left(\mathcal{M}^\perp\right)^\perp = \overline{\mathcal{M}}\,. \tag{2.13}$$

Proof Note first that $\mathcal{M} \subset \left(\mathcal{M}^\perp\right)^\perp$ (if $h \in \mathcal{M}$, then $(h,x) = 0\ \forall x \in \mathcal{M}^\perp$, hence $h \in \left(\mathcal{M}^\perp\right)^\perp$). Taking closures we obtain $\overline{\mathcal{M}} \subset \left(\mathcal{M}^\perp\right)^\perp$.

Next, let us take $x \in \left(\mathcal{M}^\perp\right)^\perp$. We can write: $x = y + z$, where $y \in \overline{\mathcal{M}}$, $z \in \overline{\mathcal{M}}^\perp = \mathcal{M}^\perp$. As $x \perp \mathcal{M}^\perp$, we get $(x,z) = 0$. Thus $0 = (y+z,z) = (y,z) + \|z\|^2 = \|z\|^2$; hence $z = \mathbf{0}$, and $x = y \in \overline{\mathcal{M}}$. ■

3. Normed linear spaces and linear operators

We complete here the previous discussion of normed linear spaces. Thus, a sequence $(x_n)_1^\infty$ in the normed linear space E is said to be convergent to the element $x_0 \in E$ iff $\|x_n - x_0\| \to 0$ as $n \to \infty$. Next, a sequence $(x_n)_1^\infty$ in E is a Cauchy sequence iff $\forall \varepsilon > 0\ \exists \bar{k} \in \mathbb{N}$ such that $\|x_n - x_m\| < \varepsilon$ whenever $m \ge \bar{k}$ and $n \ge \bar{k}$.

If every Cauchy sequence in E is convergent to some element in E, we say that E is a Banach space ('complete normed space').

An open ball $B(x_0,\rho)$ in E (having center x_0 and radius ρ) is given by the inequality:

$$B(x_0,\rho)=\{x\in E,\ \|x-x_0\|<\rho\} \tag{3.1}$$

while the closed ball $C(x_0,\rho)$ (again with center x_0 and radius ρ) will be defined by the inequality

$$C(x_0,\rho)=\{x\in E,\ \|x-x_0\|\le\rho\}\ . \tag{3.2}$$

Next, it is possible to define a bounded set A in the normed linear space E by the following property:

$A\subset E$ is bounded iff $\exists$ a ball $B(\mathbf{0},\rho)$ such that $x\in B(\mathbf{0},\rho)\quad \forall x\in A$.

(Equivalently, if $\exists\rho>0$ such that $\|x\|<\rho\ \forall x\in A$.)

Consider the next two linear normed spaces over the same field K; let E and F be these two spaces. Next, consider T, a linear mapping (function or operator) from E (that is with domain E) into F (with range in F).

Linearity of the mapping T means, as usual, that

$$T(\alpha x+\beta y)=\alpha T(x)+\beta T(y)\ ,\quad \forall\alpha,\beta\in K,\quad \forall x,y\in E\ . \tag{3.3}$$

We say that T is a continuous mapping if the following holds:

$$(x_n)_1^\infty \ \text{ in } E \quad\text{and}\quad x_n\to x_0\in E\Rightarrow Tx_n\to Tx_0 \ \text{ in } F\ . \tag{3.4}$$

The following is of use:

Theorem 3.1 *Any linear mapping T: $E\to F$ is continuous iff, $\forall$ bounded sets $A\subset E$, the image set $T(A)$ is bounded in F.*

Proof

(1) Let us assume that T is continuous and that A is a bounded set in E. If $T(A)$ is *not* bounded in F we can find, $\forall n\in\mathbb{N}$, elements $x_n\in A$ such that $\|Tx_n\|_F\ge n$. It follows

that ${x_n}/{n} \to \mathbf{0}$ in E, while $\left\| T\left({x_n}/{n}\right)\right\|_F = \frac{1}{n}\|Tx_n\|_F \geq 1$; this contradicts the continuity of T in the origin of E.

(2) Conversely, we note that if $\|x\|_E < 2$, then $\|Tx\|_F < C$ (for some $C > 0$). Therefore, $\forall x \in E$, $x \neq \mathbf{0}$, we get $\left\| T\left({x}/{\|x\|_E}\right)\right\|_F < C$, hence $\|Tx\|_F < C\|x\|_E$ and then

$$\|T(x-y)\|_F \leq C\|x-y\|_F \,, \quad \forall x, y \in E \tag{3.5}$$

which says that T is a Lipschitz continuous function on E. ■

We now consider a particular - but fundamental - case, where the second vector space F reduces to the field K (viewed as a vector space).

Thus, we introduce the dual space to E, here denoted with E', which is composed of all the linear continuous mappings from E into K. We note that E' is in itself a linear-vector-space over K, the space $\mathcal{L}(E, K)$, with the usually defined addition and scalar multiplication.

We now define on E' a function: $f \in E' \to \|f\| \in [0, \infty)$, by the formula

$$\|f\|_{E'} = \sup_{\|x\|_E \leq 1} |f(x)| \,. \tag{3.6}$$

(We shall see later - in a more general situation - that the notation $\|f\|_{E'}$ is justified; it is indeed a norm on E'.)

Also, it will be seen later - again in a more general framework - that the vector space E' endowed with the norm (3.6) is a *complete* normed space (hence a Banach space).

At this stage, however, we restrict ourselves to Hilbert spaces and shall give a fundamental result (the 'F. Riesz representation theorem') in the form of:

Theorem 3.2 *Let $\mathcal{H}$ be a Hilbert space and let $\mathcal{H}' = \mathcal{L}(\mathcal{H}; \mathbb{C})$ be its dual space. Then, for any $F \in \mathcal{H}'$ there exists one and only one element $y_F \in \mathcal{H}$, such that the representation formula*

$$F(x) = (x, y_F) \quad \forall x \in \mathcal{H} \tag{3.7}$$

holds true. Furthermore, we have the equality of norms

$$\|F\|_{\mathcal{H}'} = \|y_F\|_{\mathcal{H}} \cdot \tag{3.8}$$

Remark Any element $y \in \mathcal{H}$ determines an element F_y in $\mathcal{H}'$ by means of the formula

$$F_y(x) = (x, y) \tag{3.9}$$

(it is easy to check this property).

Proof of Theorem 3.2 Let us consider the so-called kernel space of F:

$$\text{Ker } F = \{x \in \mathcal{H},\ F(x) = 0\} \ . \tag{3.10}$$

'Obviously', Ker F is a linear closed subspace of $\mathcal{H}$. If Ker $F = \mathcal{H}$ we can take $y_F = \mathbf{0}$. Otherwise, Ker F will be a proper subspace of $\mathcal{H}$ which we denote by M. It will follow that there exists an element $y_0 \neq \mathbf{0},\ y_0 \perp M$. (In fact, as M is a proper subspace, $\exists x \in \mathcal{H},\ x \notin M$. Now use Theorem 2.1 and get $x = x_1 + x_2$, where $x_1 \in M,\ x_2 \in M^{\perp}$. As $x \notin M$ it follows that $x_2 \neq \mathbf{0}$; thus we can take $y_0 = x_2$.)

Next we show that, for some $\alpha \in \mathbb{C}$, we have

$$F(x) = (x, \alpha y_0) \quad \forall x \in \mathcal{H} \ . \tag{3.11}$$

Note first that if $x \in M$, (3.11) holds for all α.

If $x = y_0$ we obtain

$$F(y_0) = \overline{\alpha}\|y_0\|^2 \quad \text{and} \quad \overline{\alpha} = \frac{F(y_0)}{\|y_0\|^2} \ .$$

Thus, the number α has already been found. However, we must ensure that with this choice of α equality (3.11) is valid $\forall x \in \mathcal{H}$. As $y_0 \notin M$, $F(y_0) \neq 0$. We have, $\forall x \in \mathcal{H}$, the representation formula

$$x = \frac{F(x)}{F(y_0)} y_0 + \left(x - \frac{F(x)}{F(y_0)} y_0 \right) . \tag{3.12}$$

Furthermore, we see that $F\left(x-\frac{F(x)}{F(y_0)}y_0\right)=0$, hence $x-\frac{F(x)}{F(y_0)}y_0 \in M$. Then

$$(x,\alpha y_0)=\left(\frac{F(x)}{F(y_0)}y_0,\ \alpha y_0\right)=\overline{\alpha}\frac{F(x)}{F(y_0)}\|y_0\|^2=F(x)\ .$$

This proves the 'existence' part with $y_F=\alpha y_0$.

The uniqueness of y_F in (3.7) is immediate, for, if $(x,y_F^1)=(x,y_F^2)$, $\forall x\in\mathcal{H}$, we get, taking $x=y_F^1-y_F^2$, that $\|y_F^1-y_F^2\|=0$.

Finally, let us establish (3.8); first we have

$$\|F\|_{\mathcal{H}'}=\sup_{\|x\|\le 1}|F(x)|=\sup_{\|x\|\le 1}|(x,y_F)|\le\sup_{\|x\|\le 1}(\|x\|\,\|y_F\|)=\|y_F\|\ ,$$

and then

$$\|F\|_{\mathcal{H}'}\ge\left|F\left(\frac{y_F}{\|y_F\|}\right)\right|=\left(\frac{y_F}{\|y_F\|},\ y_F\right)=\|y_F\|\ .$$

■

A corollary of the above result, quite important for some applications, concerns a class of 'sesquilinear forms' on the cartesian product $\mathcal{H}\times\mathcal{H}$ where $\mathcal{H}$ is a Hilbert space. Precisely, if $\mathcal{H}$ is a Hilbert space over K, a function $(x,y)\to B(x,y)$ from $\mathcal{H}\times\mathcal{H}$ into K is a sesquilinear form if it satisfies the following properties:

(i) $B(\alpha x+\beta y,z)=\alpha B(x,z)+\beta B(y,z),\quad \forall\alpha,\beta\in K,\ x,y,z\in\mathcal{H}$ (3.13)

(ii) $B(x,\alpha y+\beta z)=\overline{\alpha}B(x,y)+\overline{\beta}B(x,z),\quad \forall\alpha,\beta\in K,\ x,y,z\in\mathcal{H}.$ (3.14)

We next assume also the property:

(iii) $|B(x,y)|\le C\|x\|\,\|y\|\quad \forall x,y\in\mathcal{H}$ (3.15)

and then we are ready for the statement of the following:

Theorem 3.3 *Under the above assumptions (i)-(iii) there exists one and only one linear continuous operator* $A:\mathcal{H}\to\mathcal{H}$, *such that*

$$B(x,y)=(Ax,y)\ ,\quad \forall x,y\in\mathcal{H}\ . \tag{3.16}$$

Moreover, the equality

$$\tilde{C} = \inf\{C > 0,\ |B(x,y)| \le C\|x\|\,\|y\|,\ \forall x, y \in \mathcal{H}\} = \sup_{\|x\|\le 1} \|Ax\|_{\mathcal{H}} \tag{3.17}$$

holds true.

Remark The expression $\sup_{\|x\|\le 1} \|Ax\|_{\mathcal{H}}$ is, by definition, the norm of the operator A. We shall return to this definition later.

Proof of Theorem 3.3

(a) First we show uniqueness in representation (3.16). Thus, if $B(x,y) = (A_1 x, y) = (A_2 x, y)$, $\forall x, y \in \mathcal{H}$, we obtain $((A_1 - A_2)x, y) = 0,\ \forall x, y \in \mathcal{H}$. Take $y = (A_1 - A_2)x$ and we get $\|(A_1 - A_2)x\|^2 = 0$, hence

$$A_1 x = A_2 x\,,\quad \forall x \in \mathcal{H}\,,\quad A_1 = A_2\ .$$

(b) Next we establish (3.17). From (3.16) we derive $|(Ax, y)| \le C\|x\|\,\|y\|,\ \forall x, y \in \mathcal{H}$. Take then $y = Ax$ and one obtains

$$\|Ax\|^2 \le C\|x\|\,\|Ax\|,\quad \forall x \in \mathcal{H},\quad \text{hence}\quad \|Ax\| \le C\|x\|\,,\ \forall x \in \mathcal{H}\ .$$

Therefore we see that $\sup_{\|x\|\le 1} \|Ax\| \le C$ for all C appearing in (3.17), hence also

$$\sup_{\|x\|\le 1} \|Ax\| \le \tilde{C}\ . \tag{3.18}$$

Next, if one puts - as indicated in the Remark above - $\|A\| = \sup_{\|x\|\le 1} \|Ax\|$, we first see that, $\forall y \in \mathcal{H},\ y \ne \mathbf{0}$, the inequality

$$\|Ay\| = \left\| A \frac{y}{\|y\|} \right\| \|y\| \le \|A\|\,\|y\| \tag{3.19}$$

holds. (It holds obviously for $y = \mathbf{0}$ too!)

Now, turning back to (3.17), we find

$$|B(x,y)| = |(Ax, y)| \le \|Ax\|\,\|y\| \le \|A\|\,\|x\|\,\|y\|\,,\qquad \forall x, y \in \mathcal{H}\,, \tag{3.20}$$

therefore $\tilde{C} \le \|A\|$, which, together with (3.18), will give $\tilde{C} = \|A\|$.

Finally, we arrive at the existence part in (3.16). Let us fix $x \in \mathcal{H}$ and then consider the mapping

$$y \in \mathcal{H} \to \overline{B(x,y)} \in \mathbb{C} \text{ or } \mathbb{R} . \tag{3.21}$$

It is easily seen to be a linear form (functional) on $\mathcal{H}$. Furthermore, we have the estimate

$$\left|\overline{B(x,y)}\right| \le C\|x\|\,\|y\|$$

which indicates that the functional in (3.21) is also a continuous functional on $\mathcal{H}$. Hence, we are entitled to apply Theorem 3.2; one finds a unique element $w_x \in \mathcal{H}$, such that

$$\overline{B(x,y)} = (y, w_x) \quad \forall y \in \mathcal{H},\ B(x,y) = (w_x, y),\ \forall y \in \mathcal{H} . \tag{3.22}$$

We now put $Ax = w_x$, hence $B(x,y) = (Ax, y)$, $\forall y \in \mathcal{H}$ (and then, $\forall x \in \mathcal{H}$). We note that A is a linear operator: we have in fact

$$B(x_1 + x_2, y) = B(x_1, y) + B(x_2, y) = (Ax_1, y) + (Ax_2, y) = (Ax_1 + Ax_2, y)$$

and also $B(x_1 + x_2, y) = (A(x_1 + x_2), y)$. Thus we get the equality

$$(A(x_1 + x_2), y) = (Ax_1 + Ax_2, y) , \quad \forall y \in \mathcal{H}$$

which obviously entails

$$A(x_1 + x_2) = Ax_1 + Ax_2 , \quad \forall x_1, x_2 \in \mathcal{H} .$$

In a similar way one obtains the equality

$$A(\alpha x) = \alpha Ax , \quad \forall \alpha \in K,\ \forall x \in \mathcal{H} .$$

Finally, in order to prove the continuity of operator A, we use the estimates

$$|(Ax,y)| = |B(x,y)| \le C\|x\|\,\|y\|\,, \qquad \forall x,y \in \mathcal{H}\ .$$

If $y = Ax$ we derive

$$\|Ax\|^2 \le C\|x\|\,\|Ax\|\,, \qquad \text{hence} \qquad \|Ax\| \le C\|x\|\,, \qquad \forall x \in \mathcal{H}$$

(this last deduction already occurred in (b)). ■

Remark Let $A \in \mathcal{L}(\mathcal{H},\mathcal{H}) = \mathcal{L}(\mathcal{H})$ (a linear continuous operator $\mathcal{H} \to \mathcal{H}$). Then the function B defined on $\mathcal{H} \times \mathcal{H}$ through the formula

$$B(x,y) = (Ax,y)$$

verifies all the conditions (3.13)-(3.15).

We shall now end this section with

Theorem 3.4 *(The dual of a Hilbert space is also a Hilbert space) Let H be a Hilbert space and let H' be its dual space. Then H' can be given a Hilbert space structure and there is an isometric conjugate-linear mapping σ from H' onto H.*

Proof Remember that H' is the Banach space $\mathcal{L}(H;K)$, where $\|f\|_{H'} = \sup\limits_{\|h\|\le 1}|f(h)|$. As expected, the essential ingredient in the proof will be the representation Theorem 3.2. Thus, $\forall f \in H'$, $\exists!\, y_f \in H$ in such a way that

$$f(x) = \left(x, y_f\right), \qquad \forall x \in H\ . \tag{3.23}$$

We shall accordingly define a scalar product on H' by the formula

$$(f,g)_{H'} = \overline{\left(y_f, y_g\right)}_H\,, \qquad \forall f,g \in H'\ . \tag{3.24}$$

It is indeed a scalar product: $(f,f)_{H'} = \left\|y_f\right\|^2 \ge 0$. If $(f,f)_{H'} = 0$, then $y_f = \mathbf{0}$, and hence, $f = \mathbf{0}$ in H'.

Next, for f, g_1, g_2 in H' we have $(f, g_1 + g_2)_{H'} = \overline{(y_f, y_{g_1+g_2})}_H$; on the other hand, we see that $y_{g_1+g_2} = y_{g_1} + y_{g_2}$:

$$(g_1 + g_2)(x) = g_1(x) + g_2(x) ;$$

hence

$$\left(x, y_{g_1+g_2}\right) = \left(x, y_{g_1}\right) + \left(x, y_{g_2}\right) = \left(x, y_{g_1} + y_{g_2}\right), \quad \forall x \in H, \tag{3.25}$$

and accordingly, $y_{g_1+g_2} = y_{g_1} + y_{g_2}$.

We obtain this way:

$$(f, g_1 + g_2)_{H'} = \overline{\left(y_f, y_{g_1} + y_{g_2}\right)} = \overline{\left(y_f, y_{g_1}\right)} + \overline{\left(y_f, y_{g_2}\right)} = (f, g_1)_{H'} + (f, g_2)_{H'} .$$

We next see that $(f, \lambda g)_{H'} = \overline{\lambda}(f, g)_{H'}$, $\forall \lambda \in K$, $f, g \in H'$.

In fact, $(f, \lambda g)_{H'} = \overline{\left(y_f, y_{\lambda g}\right)}_{H'}$ and now $y_{\lambda g} = \overline{\lambda} y_g$; this is because of the following:

$$(\lambda g)(x) = \left(x, y_{\lambda g}\right) = \lambda g(x) = \lambda\left(x, y_g\right) = \left(x, \overline{\lambda} y_g\right), \quad \forall x \in H .$$

Therefore:

$$(f, \lambda g)_{H'} = \overline{\left(y_f, \overline{\lambda} y_g\right)} = \overline{\lambda}\,\overline{\left(y_f, y_g\right)} = \overline{\lambda}(f, g)_{H'} .$$

Finally

$$(f, g)_{H'} = \overline{\left(y_f, y_g\right)}_H = \left(y_g, y_f\right)_H = \overline{(g, f)}_{H'} .$$

Now, we also know that

$$\|f\|_{H'} = \sup_{\|x\| \le 1} \left|\left(x, y_f\right)_H\right| = \left\|y_f\right\|_H = (f, f)_{H'}^{\frac{1}{2}} .$$

The norm induced by the scalar product in H' is the norm in H' as a dual space. Hence, the (pre-) Hilbert space H' is also complete.

Let us now define the mapping σ, from H' into H, by the relation

$$\sigma(f) = y_f , \quad \forall f \in H' . \tag{3.26}$$

First we have

$$\|\sigma(f)\|_H = \|y_f\|_H = \|f\|_{H'}, \qquad \forall f \in H'.$$

We say that σ is an isometric mapping. Also, σ is onto ('surjective' mapping): for, if $y \in H$, we define f by $f(x) = (x, y)$ and so $\sigma(f) = y$.

Next σ is an 'additive' mapping:

$$\sigma(f_1 + f_2) = y_{f_1+f_2} = y_{f_1} + y_{f_2} = \sigma(f_1) + \sigma(f_2).$$

However, we only have $\sigma(\lambda f) = \bar{\lambda}\sigma(f)$; this is why we say that σ is conjugate linear (indeed, $\sigma(\lambda f) = y_{\lambda f} = \bar{\lambda} y_f = \bar{\lambda}\sigma(f)$).

Note, finally, that

$$(\sigma(f), \sigma(g))_H = (y_f, y_g)_H = \overline{(f, g)}_{H'} = (g, f)_{H'}. \tag{3.27}$$

■

4. Orthonormal bases

We start this section with a statement of 'Zorn's Lemma', as usually found in mathematics books. But first we give a definition: a relation on a set X (here denoted by R), which is reflexive (xRx), transitive (xRy and $yRz \Rightarrow xRz$), and antisymmetric (xRy and $yRx \Rightarrow x = y$) is called a *partial* ordering (some couples in X may not be related at all!). Sometimes we write $x \prec y$ instead of xRy.

Example Let $\mathcal{P}(X)$ be the collection of all the subsets of a set X. Define, for $A, B \in \mathcal{P}(X)$, the relation $A \prec B$ iff $A \subset B$. Then (as is easily seen), we have a partial ordering.

Furthermore, let us assume that $\forall x, y \in X$, either $x \prec y$ or $y \prec x$. We say in this situation that X is totally (or linearly) ordered.

Definition 4.1 *Let Y be a subset of the partially ordered set X. An element $p \in X$ is an upper bound for Y if $y \prec p \ \forall y \in Y$. An element $m \in X$ is a maximal element of X if $m \prec x \Rightarrow x = m$ (where $x \in X$).*

We now state (without proof!):

Theorem 4.1 *Let X be a (non-empty) partially ordered set with the property that every totally ordered subset has an upper bound in X. Then X has a maximal element.*

(This is 'Zorn's Lemma'.)

We shall now give:

Definition 4.2 *Let S be an orthonormal set in the Hilbert space $\mathcal{H}$ which is not strictly contained in any other orthonormal set in $\mathcal{H}$. We say that S is an orthonormal basis, or also a complete orthonormal system for $\mathcal{H}$.*

As a quite simple application of Theorem 4.1, we shall now state an existence result for orthonormal bases.

Theorem 4.2 *Every Hilbert space $\mathcal{H}$ containing a non-zero element possesses an orthonormal basis.*

Proof Let us consider the collection $\mathcal{C}$ of all orthonormal systems in $\mathcal{H}$, ordered by set inclusion. That is, if S_1, S_2 are in $\mathcal{C}$, we say that $S_1 \prec S_2$ whenever $S_1 \subset S_2$.

Note first that $\mathcal{C}$ is not empty, for, if $h \in \mathcal{H}$, $h \neq \mathbf{0}$, then the singleton $\left\{ h/\|h\| \right\}$ is an orthonormal system.

Next, let $\{S_\alpha\}$ be a totally ordered subset of $\mathcal{C}$. It follows that $\bigcup S_\alpha$ is also an orthonormal set in $\mathcal{H}$: $h \in \bigcup S_\alpha \Rightarrow h \in S_{\alpha_0}$ for some α_0, hence $\|h\| = 1$; furthermore, if $h \in \bigcup S_\alpha$ and $k \in \bigcup S_\alpha$, then $h \in S_{\alpha_1}$ while $k \in S_{\alpha_2}$; if, say, $S_{\alpha_1} \subset S_{\alpha_2}$, we get that both h and k are in S_{α_2}, which entails that $h \perp k$ (provided $h \neq h$!). Finally, each S_α is contained in $\bigcup S_\alpha$, which now appears as an upper bound for $\{S_\alpha\}$.

We now apply Theorem 4.1 and find in $\mathcal{C}$ a maximal element, that is an orthonormal system in $\mathcal{H}$, which is not strictly contained in any other orthonormal set in $\mathcal{H}$, that is, an orthonormal basis in $\mathcal{H}$. ■

The following is a fundamental result:

Theorem 4.3 *Let $\mathcal{H}$ be a Hilbert space and let $S=\{x_\alpha\}$ be an orthonormal basis of $\mathcal{H}$. Then the following holds:*

(i) $\forall y \in \mathcal{H}$, *we have* $(y, x_\alpha)=0$ *for all but a countable set* $(\alpha_j)_1^\infty$ *depending on* y;

(ii) *the representation formula*

$$y=\sum_{j=1}^{\infty}(y, x_{\alpha_j})x_{\alpha_j}$$

is true (in the sense that

$$\lim_{N\to\infty}\left\| y-\sum_{j=1}^{N}(y, x_{\alpha_j})x_{\alpha_j}\right\| = 0\ ); \tag{4.1}$$

(iii) *the equality*

$$\|y\|^2=\sum_{j=1}^{\infty}\left|(y, x_{\alpha_j})\right|^2 \tag{4.2}$$

is satisfied.

Proof We first use Bessel's inequality (1.4). For any finite subset of S: $\{x_\alpha, \alpha\in F\}$, where F is a finite subset of the index set A of the α's; we have

$$\sum_{\alpha\in F}\left|(y, x_\alpha)\right|^2 \le \|y\|^2 . \tag{4.3}$$

Define now, $\forall k \in \mathbb{N}$, the set of indexes

$$A_k=\left\{\alpha\in A,\ \left|(y, x_\alpha)\right|>\frac{1}{k}\right\} . \tag{4.4}$$

We see that each A_k is a finite subset of A; the number of its elements being $\le k^2\|y\|^2$ (otherwise, if A_k has more than $k^2\|y\|^2$ elements, we get

$$\sum_{\alpha \in A_k'} \left|(y, x_\alpha)\right|^2 > \frac{1}{k^2} \cdot k^2 \|y\|^2 = \|y\|^2 ,$$

in contradiction with (4.3) where A_k' is any finite subset of A_k with more than $k^2\|y\|^2$ elements).

Next, note that

$$\left\{\alpha \in A,\ \left|(y, x_\alpha)\right| > 0\right\} \subset \bigcup_{k=1}^{\infty} A_k . \tag{4.5}$$

Thus, the set $\left\{\alpha \in A,\ (y, x_\alpha) \neq 0\right\}$ is countable[1]. Let us denote it by A_y. The assertion (i) is established.

Let us consider now, $\forall n \in \mathbb{N}$, elements $y_n = \sum_{j=1}^{n} \left(y, x_{\alpha_j}\right) x_{\alpha_j}$. We prove that the sequence $(y_n)_1^\infty$ is a Cauchy sequence in $\mathcal{H}$. We first note that, for $n > m$,

$$\|y_n - y_m\|^2 = \left\| \sum_{j=m+1}^{n} \left(y, x_{\alpha_j}\right) x_{\alpha_j} \right\|^2 = \sum_{j=m+1}^{n} \left|\left(y, x_{\alpha_j}\right)\right|^2 . \tag{4.6}$$

On the other hand, the numerical sequence $(\sigma_n)_1^\infty$, where $\sigma_n = \sum_{k=1}^{n} \left|\left(y, x_{\alpha_k}\right)\right|^2$, is monotone increasing and also (in view of (4.3)), bounded above by $\|y\|^2$. Thus, (σ_n) is a Cauchy sequence, and this fact, introduced in (4.6), shows that $(y_n)_1^\infty$ is a Cauchy sequence in $\mathcal{H}$. Let $\bar{y} \in \mathcal{H}$, $\bar{y} = \lim y_n$.

We note that

$$\left(y - \bar{y}, x_{\alpha_l}\right) = \lim_{n \to \infty} \left(y - \sum_{j=1}^{n} \left(y, x_{\alpha_j}\right) x_{\alpha_j}, x_{\alpha_l} \right) = 0 , \qquad \forall l = 1, 2, . \tag{4.7}$$

Next, if $\alpha \in A / A_y$ we get $(y, x_\alpha) = 0$ and then

$$(\bar{y}, x_\alpha) = \lim_{n \to \infty} \left(\sum_{j=1}^{n} \left(y, x_{\alpha_j}\right) x_{\alpha_j}, x_\alpha \right) = 0 . \tag{4.8}$$

[1] i.e., can be put in one-one correspondence with a subset of $\mathbb{N}$.

We thus obtain that $y-\bar{y} \perp S$.

If $y-\bar{y} \neq \mathbf{0}$, the set $S \cup \left\{ \dfrac{y-\bar{y}}{\|y-\bar{y}\|} \right\}$ is orthonormal and strictly $\succ S$, contradicting the maximality of S.

Therefore we get $y = \bar{y} = \lim\limits_{n\to\infty} \sum\limits_{j=1}^{n} \left(y, x_{\alpha_j}\right) x_{\alpha_j}$, which is (4.1).

Finally, we also have

$$0 = \lim_{n\to\infty} \left\| y - \sum_{j=1}^{n} \left(y, x_{\alpha_j}\right) x_{\alpha_j} \right\|^2$$

$$= \lim_{n\to\infty} \left(y - \sum_{j=1}^{n} \left(y, x_{\alpha_j}\right) x_{\alpha_j},\ y - \sum_{j=1}^{n} \left(y, x_{\alpha_j}\right) x_{\alpha_j} \right)$$

$$= \lim_{n\to\infty} \left\{ \|y\|^2 - \sum_{j=1}^{n} \left|\left(y, x_{\alpha_j}\right)\right|^2 - \sum_{j=1}^{n} \left|\left(y, x_{\alpha_j}\right)\right|^2 + \sum_{j=1}^{n} \left|\left(y, x_{\alpha_j}\right)\right|^2 \right\}$$

$$= \lim_{n\to\infty} \left\{ \|y\|^2 - \sum_{j=1}^{n} \left|\left(y, x_{\alpha_j}\right)\right|^2 \right\} \tag{4.9}$$

which is (4.2). ■

In order to proceed to the next result, let us remember the following:

Definition 4.3 *A linear normed space is called separable if it contains a countable dense subset.*

We now state:

Theorem 4.4 *Any separable Hilbert space possesses a countable orthonormal basis.*

Proof Note first that if $\{x_\alpha\}$ is any orthonormal system in the Hilbert space $\mathcal{H}$, we have

$$\left\| x_\alpha - x_\beta \right\|^2 = \left(x_\alpha - x_\beta,\ x_\alpha - x_\beta \right) = 2 \quad \text{whenever } \alpha \neq \beta\ . \tag{4.10}$$

Therefore, the balls in $\mathcal{H}$, $B_\alpha = \left\{x \in \mathcal{H},\ \|x - x_\alpha\| < \frac{1}{2}\right\}$, are mutually disjoint.

Next, we previously proved that $\mathcal{H}$ has an orthonormal basis; let it be $\{x_\alpha\}$.

Furthermore, by assumption, $\mathcal{H}$ also has a countable dense subset $Y = \{y_k\}_{k \in \mathbb{N}}$. Thus $\overline{Y} = \mathcal{H}$, and each ball B_α contains some y_k [2].

Also, if $\alpha \neq \beta$, the balls B_α and B_β contain different y_k. (If a ball B_α contains more than one y_g, we take just one of them.) We have in this way an injective mapping from the balls B_α to a subset of $\{y_k\}$ (onto this subset in fact).

Hence, the collection of balls $\{B_\alpha\}$ is countable. As it is in one-one correspondence with the basis $\{x_\alpha\}$, this basis is also countable. ■

5. Some additional results in Hilbert space

Proposition 5.1 *Let $(x_n)_1^\infty$ be an orthonormal sequence in the pre-Hilbert space V. Then:* $\lim\limits_{n\to\infty}(x_n, h) = 0\ \forall h \in V$.

Note. The sequence (x_n) in V is orthonormal iff the set $\{x_n\}$ is orthonormal.

Now, from Proposition 1.1 we find that $\sum\limits_{n=1}^{N} |(x_n, h)|^2 \le \|h\|^2$, $\forall N \in \mathbb{N}$, hence $\sum\limits_{n=1}^{\infty} |(x_n, h)|^2 < \infty$; therefore $|(x_n, h)|^2 \to 0$ and then $|(x_n, h)| \to 0$ (as $n \to \infty$). ■

Next, let us give:

Proposition 5.2 *Let V be a pre-Hilbert space and let $\{x_n\}$ $(n = 1, 2, \ldots, N)$ be a finite orthonormal set in V. Then the expression*

$$\left\| x - \sum_{n=1}^{N} c_n x_n \right\|, \qquad x \in V, \tag{5.1}$$

is minimal for $c_n = (x, x_n)$.

We have in fact the following:

---

[2] For instance, if $(x_j) \to x_\alpha$ and $\{x_j\} \subset Y$, one can take the first $x_{\bar{j}}$ belonging to B_α.

$$\left\| x - \sum_{n=1}^{N} c_n x_n \right\|^2 = \left(x - \sum_{1}^{N} c_n x_n,\ x - \sum_{1}^{N} c_n x_n \right)$$

$$= \|x\|^2 - \sum_{1}^{N} c_n (x_n, x) - \sum_{1}^{N} (x, c_n x_n) + \sum_{1}^{N} |c_n|^2$$

$$= \|x\|^2 + \sum_{1}^{N} \left[c_n - (x, x_n)\right]\left[\bar{c}_n - (x_n, x)\right] - \sum_{1}^{N} |(x, x_n)|^2$$

$$= \|x\|^2 + \sum_{1}^{N} |c_n - (x, x_n)|^2 - \sum_{1}^{N} |(x, x_n)|^2$$

$$\geq \|x\|^2 - \sum_{1}^{N} |(x, x_n)|^2 . \tag{5.2}$$

Looking at (5.2) we see that the real-valued function

$$\phi(c_1, c_2, \ldots, c_n) = \left\| x - \sum_{1}^{N} c_n x_n \right\|^2$$

is bounded from below by $\|x\|^2 - \sum_{1}^{N} |(x, x_n)|^2$ and that this lower bound is actually attained for $c_n = (x, x_n)$. ■

The following statement (concerning the unique extension of linear continuous functionals on a linear subset of a Hilbert space to the whole space), is quite interesting.

Theorem 5.1 *Let $\mathcal{M}$ be a linear subset of the Hilbert space $\mathcal{H}$. Let $f: \mathcal{M} \to \mathbb{C}$ be a linear function, such that*

$$\sup |f(x)|, \quad x \in \mathcal{M}, \quad \|x\| \leq 1 = C . \tag{5.3}$$

Then, there exists a unique extension of f, F, a linear function $\mathcal{H} \to \mathbb{C}$ such that

$$\sup |F(h)|, \quad h \in \mathcal{H}, \quad \|h\| \leq 1 = C . \tag{5.4}$$

(With the notation (3.6) we assume $\|f\|_{\mathcal{M}'} = C$ and find unique $F \in \mathcal{H}'$, $F = f$ on $\mathcal{M}$, and $\|F\|_{\mathcal{H}'} = C$.)

Note. The existence of the linear extension F appears as a particular case of the so-called Hahn-Banach theorems. However, the proof given below is different.

Proof We first consider the closure $\overline{\mathcal{M}}$ of the linear subset $\mathcal{M}$ in $\mathcal{H}$. Thus we now have a closed linear subspace $\overline{\mathcal{M}}$ of $\mathcal{H}$.

The extension of f from $\mathcal{M}$ to $\overline{\mathcal{M}}$ is now effected in the standard way: as seen in (3.5), f is Lipschitz-continuous on $\mathcal{M}$:

$$|f(x) - f(y)| = |f(x-y)| \leq c\|x-y\| , \qquad \forall x, y \in \mathcal{M} . \tag{5.5}$$

Therefore, f is uniformly continuous on $\mathcal{M}$. Now, if $x \in \overline{\mathcal{M}}$, there exists a sequence (x_n) in $\mathcal{M}$, where $x_n \to x$. Then the numerical sequence $(f(x_n))$ is a Cauchy sequence (in view of (5.5)), hence $\lim\limits_{n\to\infty} f(x_n)$ exists. Furthermore, this limit is independent of the particular sequence (x_n) which converges to x, as can be seen by consideration of the alternate sequence x_1, x_1', x_2, $x_2', \ldots$ formed with (x_n) and another sequence (x_n') in $\mathcal{M}$, where $x_n' \to x$ too.

Thus, we can define $f(x) = \lim\limits_{n\to\infty} f(x_n)$. We have by now an extension of f, from $\mathcal{M}$ to $\overline{\mathcal{M}}$.

Actually, f is also linear on $\overline{\mathcal{M}}$ (for instance, if $x_1, x_2 \in \overline{\mathcal{M}}$, and $x_{1,n} \to x_1$, $x_{2,n} \to x_2$, we get: $f(x_1) = \lim f(x_{1,n})$, $f(x_2) = \lim f(x_{2,n})$, $f(x_1) + f(x_2) = \lim\limits_{n\to\infty}\left[f(x_{1,n}) + f(x_{2,n})\right]$ $= \lim\limits_{n\to\infty} f(x_{1,n} + x_{2,n}) = f(x_1 + x_2)$).

Note also that the 'norm of f' on $\overline{\mathcal{M}}$, is the number

$$\sup|f(x)| , \qquad x \in \overline{\mathcal{M}}, \qquad \|x\| \leq 1 \tag{5.6}$$

that equals the norm of f on $\mathcal{M}$ given in (5.3). (In fact, first of all we have the obvious inequality

$$\sup_{\substack{x \in \overline{\mathcal{M}} \\ \|x\| \leq 1}} |f(x)| \geq \sup_{\substack{x \in \mathcal{M} \\ \|x\| \leq 1}} |f(x)| = \|f\|_{\mathcal{M}'} = C ;$$

next, let $x \in \overline{\mathcal{M}}$, $\|x\| \le 1$, $x \ne \mathbf{0}$. Take (x_n) in $\mathcal{M}$, $x_n \to x$, $x_n \ne \mathbf{0}$. Then $x_n / \|x_n\| \to x / \|x\|$ and

$$|f(x)| = \lim |f(x_n)| = \lim \left| f\left(\frac{x_n}{\|x_n\|} \cdot \|x_n\| \right) \right| \le C \lim \|x_n\| = C\|x\| \le C \;;$$

thus $\sup\limits_{\substack{x \in \overline{\mathcal{M}} \\ \|x\| \le 1}} |f(x)| \le C$ and we get the required equality

$$\|f\|_{(\overline{\mathcal{M}})'} = \|f\|_{\mathcal{M}'} = C \;.) \tag{5.7}$$

Our next step consists in the extension of f from $\overline{\mathcal{M}}$ to the whole space $\mathcal{H}$. We know from previous arguments that, $\forall x \in \mathcal{H}$, there holds the unique decomposition $x = m + n$, where $m \in \overline{\mathcal{M}}$ and $n \perp \overline{\mathcal{M}}$.

We now define F, a function from $\mathcal{H}$ to $\mathbb{C}$, by

$$F(x) = f(m) \;. \tag{5.8}$$

We first see that, if $x \in \overline{\mathcal{M}}$, then $m = x$ and $F(x) = f(x)$; F is an extension of f.

It is also a linear extension; for instance, if $x = m_1 + n_1$, $y = m_2 + n_2$ then $x + y = (m_1 + m_2) + (n_1 + n_2)$ and $F(x+y) = f(m_1 + m_2) = f(m_1) + f(m_2) = F(x) + F(y)$.

Next, $F \in \mathcal{H}'$; in fact: $|F(x)| = |f(m)| \le C\|m\| \le C\|x\|$, $\forall x \in \mathcal{H}$ (note that $\|x\|^2 = \|m\|^2 + \|n\|^2$).

Furthermore, from the above, we see that $\|F\|_{\mathcal{H}'} \le C$. If $x \in \overline{\mathcal{M}}$ we have $F(x) = f(x)$ and

$$|f(x)| \le \|F\|_{\mathcal{H}'} \|x\| \,, \quad \forall x \in \overline{\mathcal{M}} \quad \text{and} \quad \|f\|_{(\overline{\mathcal{M}})'} \le \|F\|_{\mathcal{H}'} \le C \,.$$

We obtain $\|F\|_{\mathcal{H}'} = \|f\|_{\mathcal{M}'} = C$. The 'existence' part of the proof is settled.

We arrive now at the 'uniqueness' part. Let $G \in \mathcal{H}'$, such that $G = f$ on $\mathcal{M}$. Then $G = f$ on $\overline{\mathcal{M}}$.

We know also that, for a unique element $g \in \mathcal{H}$, the representation relation $G(x) = (x, g)$ holds, $\forall x \in \mathcal{H}$.

Let us write now the unique representation $g = g_1 + g_2$, $g_1 \in \overline{\mathcal{M}}$, $g_2 \in (\overline{\mathcal{M}})^{\perp}$. Then, $\forall m \in \mathcal{M}$ we obtain $G(m) = (m, g) = (m, g_1)$.

Note also that, as f extended to $\overline{\mathcal{M}}$ is in $(\overline{\mathcal{M}})'$, there exists unique $\bar{f} \in \overline{\mathcal{M}}$, such that $f(m) = (m, \bar{f})$, $\forall m \in \overline{\mathcal{M}}$.

Accordingly we obtain $G(m) = f(m) = (m, g_1) = (m, \bar{f})\ \forall m \in \overline{\mathcal{M}}$. This implies: $g_1 = \bar{f}$. On the other hand, let us remember that:

$$\|G\|_{\mathcal{H}'} = \|g\|_{\mathcal{H}}\ , \quad \|f\|_{(\overline{\mathcal{M}})'} = \|\bar{f}\|_{\mathcal{H}}\ .$$

As we look for an extension of f with conservation of the norm, we have $\|G\|_{\mathcal{H}'} = \|f\|_{\mathcal{M}'} = \|f\|_{(\overline{\mathcal{M}})'}$ and consequently we obtain $\|g\|_{\mathcal{H}} = \|\bar{f}\|_{\mathcal{H}}$. It then follows that

$$\|\bar{f}\|_{\mathcal{H}}^2 = \|g\|_{\mathcal{H}}^2 = \|g_1\|^2 + \|g_2\|^2 = \|\bar{f}\|^2 + \|g_2\|^2 \quad \text{and} \quad g_2 = \mathbf{0}\ .$$

Thus $G(x) = (x, g_1)\ \forall x \in \mathcal{H}$. Now let $x = m + n,\ m \in \overline{\mathcal{M}},\ n \perp \overline{\mathcal{M}}$; then $G(x) = (m, g_1) + (n, g_1) = (m, g_1) = f(m) = F(x)$.

We found $G = F$, the initially defined extension! ■

Chapter II

Banach spaces and linear operators

Introduction

In the previous chapter we gave some preliminary definitions and simple properties. Here we shall make a more complete study of these concepts.

1. Let X,Y be two normed linear spaces over K[3] and let $\mathcal{L}(X,Y)$ be the (linear) space of all linear continuous mappings $X \to Y$ (the sum of two linear continuous mappings $X \to Y$ obviously has the same property, and if $\lambda \in K$ while $T \in \mathcal{L}(X,Y)$, the mapping λT, $X \to Y$ which is defined by: $(\lambda T)x = \lambda(Tx)$, $\forall x \in X$, is again in $\mathcal{L}(X,Y)$.

We define the function

$$T \in \mathcal{L}(X,Y) \to \|T\| \in [0,\infty)$$

by the relation

$$\|T\| = \sup_{\|x\| \le 1} \|Tx\| \; ; \qquad \|T\| < +\infty \forall T \in \mathcal{L}(X,Y) \tag{1.1}$$

(see Theorem 3.1, Chapter I).

Note that, if $x \in X$, $x \neq \mathbf{0}$ (zero element in x), then $x/\|x\|$ has unit norm. Accordingly we get

$$\left\| T\left(x/\|x\| \right) \right\| \le \|T\| \quad \text{and} \quad \|Tx\| \le \|T\|\,\|x\| \,, \qquad \forall x \in X \,. \tag{1.2}$$

If $T = \mathbf{0}$ (the zero operator), then obviously $\|T\| = 0$. Conversely, if $\|T\|_{\mathcal{L}(X,Y)} = 0$ it follows that $Tx = \mathbf{0}$ $\forall x \in X$, hence $T = \mathbf{0}$.

---

[3] The common field of X and Y. It can be $\mathbb{R}$ or $\mathbb{C}$.

Furthermore, if $\lambda \in K$ then, for $T \in \mathcal{L}(X,Y)$ we get

$$\|\lambda T\| = \sup_{\|x\|\le 1}\|\lambda Tx\| = \sup_{\|x\|\le 1}|\lambda|\,\|Tx\| = |\lambda|\sup_{\|x\|\le 1}\|Tx\| = |\lambda|\,\|T\| \,.$$

Finally we see that, for $T_1, T_2 \in \mathcal{L}(X,Y)$,

$$\|(T_1+T_2)x\| \le \|T_1x\| + \|T_2x\| \le \|T_1\| + \|T_2\| \quad \text{if} \quad \|x\| \le 1,$$

and this entails

$$\|T_1+T_2\| \le \|T_1\| + \|T_2\| \,.$$

Thus, (1.1) defines a norm on $\mathcal{L}(X,Y)$.

Let us now point out:

Proposition 1.1 *For $T \in \mathcal{L}(X,Y)$ we have*

$$\sup_{x\ne \mathbf{0}}\frac{\|Tx\|}{\|x\|} = \inf\{C>0,\ \|Tx\| \le C\|x\|\ \forall x \in X\} = \|T\| \,. \tag{1.3}$$

Proof Using (1.2) we get, if $x \ne \mathbf{0}$,

$$\frac{\|Tx\|}{\|x\|} \le \|T\| \,,$$

hence

$$\sup_{x\ne \mathbf{0}}\frac{\|Tx\|}{\|x\|} \le \|T\| \,.$$

On the other hand, if $x \ne \mathbf{0}$ and $\|x\| \le 1$, we get

$$\|Tx\| \le \frac{\|Tx\|}{\|x\|} \le \sup_{x\ne \mathbf{0}}\frac{\|Tx\|}{\|x\|} \,;$$

hence

$$\|T\| \le \sup_{x\ne \mathbf{0}}\frac{\|Tx\|}{\|x\|} \,.$$

Again from (1.2), if $\tilde{C} = \inf\{C > 0,\ \|Tx\| \le C\|x\|\ \forall x \in X\}$, we obtain that $\tilde{C} \le \|T\|$.

Now, let us assume $\tilde{C} < \|T\|$ strictly; it follows that there exists $C > 0$, $\tilde{C} \le C < \|T\|$, and $\|Tx\| \le C\|x\|$, $\forall x \in X$. If $\|x\| \le 1$ we get $\|T\| \le C$ and then $C < \|T\| \le C$, a contradiction. (Note: from (1.2) it follows that $\|T\|$ *belongs* to the set $\{C > 0,\ \|Tx\| \le C\|x\|,\ x \in X\}$.) ∎

The following is an important result.

Theorem 1.1 *Let Y be a Banach space. Then the space $\mathcal{L}(X,Y)$ is also a Banach space.*

Note If $Y = K$ (the scalar field) we obtain that $\mathcal{L}(X,K) = X'$ (the dual space to *X*) is a Banach space - even if *X* is not complete.

Proof Let us consider a Cauchy sequence in $L(X,Y)$: $(T_n)_1^\infty$. Therefore, $\forall \varepsilon > 0$, $\exists \bar{n}(\varepsilon) \in \mathbb{N}$, such that $\|T_n - T_m\| < \varepsilon$ for $n, m \ge \bar{n}(\varepsilon)$. It follows that, $\forall x \in X$, the sequence $(T_n x)$ is a Cauchy sequence in *Y*. As *Y* is a complete space, $\lim_{n\to\infty} T_n x = y_x \in Y$ exists, $\forall x \in X$.

Now define an operator *T*, $X \to Y$, by $Tx = y_x = \lim T_n x$.

We see that $T(\alpha x_1 + \beta x_2) = \lim T_n(\alpha x_1 + \beta x_2) = \lim[\alpha T_n x_1 + \beta T_n x_2] = \alpha T x_1 + \beta T x_2$, $\forall \alpha, \beta \in K$, $\forall x_1, x_2 \in X$; *T* is accordingly a linear operator $X \to Y$.

Next, we note first that $\|Tx\| = \lim \|T_n x\|$. On the other hand, we see that $\big|\|T_n\| - \|T_m\|\big| \le \|T_n - T_m\| < \varepsilon$ if $n, m \ge \bar{n}(\varepsilon)$ which means that the numerical sequence $(\|T_n\|)$ is also a Cauchy sequence, hence a bounded sequence: $\exists C > 0$ such that $\|T_n\| \le C\ \forall n \in \mathbb{N}$. Accordingly, $\|T_n x\| \le C\|x\|\ \forall n \in \mathbb{N}$, $\forall x \in X$, and $\lim \|T_n x\| \le C\|x\|$. Thus $\|Tx\| \le C\|x\|$, $\forall x \in X$, and *T* is continuous (in fact we have $\|T_n x\| \le \|T_n\|\,\|x\|$, hence $\|Tx\| \le (\lim \|T_n\|)\|x\|$). As a final step we establish that $\|T_n - T\| \to 0$ as $n \to \infty$.

Note first that $\|T_n x - T_m x\| \le \|T_n - T_m\|\,\|x\| < \varepsilon$ if $n, m \ge \bar{n}(\varepsilon)$ and $\|x\| \le 1$. Then $\|T_n x - Tx\| = \lim_{m\to\infty} \|T_n x - T_m x\| \le \varepsilon$, if $n \ge \bar{n}(\varepsilon)$, $\|x\| \le 1$. This gives $\|T_n - T\| \le \varepsilon$ for $n \ge \bar{n}(\varepsilon)$, hence $T_n \to T$ in $\mathcal{L}(X,Y)$. ∎

Consider now three linear normed spaces over the same field *K*: X, Y, and *Z*.

Let $T \in \mathcal{L}(X,Y)$, $U \in \mathcal{L}(Y,Z)$; the product ('composition') $U \cdot T = UT$ is then well defined $X \to Z$; it obviously belongs to $\mathcal{L}(X,Z)$. Now, if $x \in X$ we have $\|(UT)x\|_Z = \|U \cdot (Tx)\|_Z \le \|U\|\,\|Tx\|_Y \le \|U\|\,\|T\|\,\|x\|_X$. This entails the inequality

$$\|U \cdot T\| \le \|U\|\,\|T\|\,. \tag{1.4}$$

Concerning the norm of an operator T in $\mathcal{L}(X,Y)$ we see that

$$\|T\| = \sup_{\|x\|=1}\|Tx\| = \sup_{\|x\|<1}\|Tx\| \,. \tag{1.5}$$

In fact, $\sup_{\|x\|=1}\|Tx\| \leq \|T\|$. Then, if $x \in X,\ x \neq \mathbf{0}$ and $\|x\| \leq 1$, we have

$$\|Tx\| = \left\| T\left(\frac{x}{\|x\|}\right) \right\| \|x\| \leq \left(\sup_{\|x\|=1}\|Tx\| \right),$$

thus

$$\sup_{\|x\|\leq 1}\|Tx\| \leq \sup_{\|x\|=1}\|Tx\| \,.$$

Next, if $\|x\| = 1$, then

$$\left\| \frac{n}{n+1}x \right\| = \frac{n}{n+1} < 1$$

and thus

$$\left\| T\left(\frac{n}{n+1}x\right) \right\| \leq \sup_{\|x\|<1}\|Tx\| \,.$$

As $n \to \infty$, $\dfrac{n}{n+1}x \to x$ and we get

$$\left\| T\left(\frac{n}{n+1}x\right) \right\| \to \|Tx\| \leq \sup_{\|x\|<1}\|Tx\| \,;$$

therefore,

$$\sup_{\|x\|=1}\|Tx\| = \|T\| \leq \sup_{\|x\|<1}\|Tx\| \leq \sup_{\|x\|\leq 1}\|Tx\| = \|T\| \,.$$

2. In this section we shall first point out two elementary properties of linear continuous functions between normed linear spaces.

Proposition 2.1 *A linear function T from a normed space X into a normed space Y is continuous iff the set* $A = \{x \in X,\ \|Tx\|_Y \leq 1\}$ *has non-empty interior.*

Proof

(a) If T is continuous then $\|Tx\| \le K$ for $\|x\| \le 1$, hence $\left\|T\left(x/K\right)\right\| \le 1$ for $\|x\| \le 1$; if $x/K = y$ we find $\|T(y)\| \le 1$ for $\|y\| \le 1/K$.

Therefore, A contains the open ball $\left\{y \in X,\ \|y\| < 1/K\right\}$.

(b) Let T be a linear function, $X \to Y$, and assume that $\left\{x,\ \|x - x_0\| < r\right\} \subset A$. Therefore $\|Tx\| \le 1$ if $\|x - x_0\| < r$. Now, if $\|x\| < r$ we obtain $\|Tx\| \le \|T(x + x_0)\| + \|Tx_0\| \le 1 + \|Tx_0\|$. Finally, for any $x \in X,\ x \ne \mathbf{0}$, we have

$$x = \frac{2}{r} \frac{x}{\|x\|} \frac{r}{2} \|x\| ,$$

where

$$\left\| \frac{x}{\|x\|} \frac{r}{2} \right\| = \frac{r}{2} < r ;$$

hence, for $x \ne \mathbf{0}$ we get

$$\|Tx\| = \frac{2}{r} \left\| T\left(\frac{x}{\|x\|} \frac{r}{2} \right) \right\| \|x\| \le \frac{2}{r}\left(1 + \|Tx_0\|\right)\|x\| \tag{2.1}$$

which obviously holds also for $x = \mathbf{0}$.

■

Proposition 2.2 *Let T be a linear function $X \to T(X) \subset Y$. Then the inverse function T^{-1}, from $T(X)$ into X, exists as a linear continuous function iff $\exists m > 0$ such that $\|Tx\| \ge m\|x\|\ \forall x \in X$.*

Before giving the proof let us remember that the function $y = Tx,\ X \to Y_1 = T(X) \subset Y$ has an inverse $x = T^{-1}y$ iff the correspondence between X and $T(X)$ is one-to-one. In this case

$$y = T\left(T^{-1}y\right), \quad \forall y \in T(X) , \qquad x = T^{-1}(Tx) , \quad \forall x \in X .$$

If T is linear it is one-to-one iff $Tx = \mathbf{0}$ iff $x = \mathbf{0}$.

A simple argument shows that the inverse of linear functions is also linear.
Now, let us give the (simple) proof.

(a) Let us assume $\|Tx\| \geq m\|x\|$, $\forall x \in X$; it follows that $Tx = \mathbf{0} \Leftrightarrow x = \mathbf{0}$, hence T^{-1} exists (with domain $T(X)$). Let $y = Tx$, $x = T^{-1}y$. We have, $\forall y \in T(X)$:

$$\|x\| = \left\|T^{-1}y\right\| \leq \frac{1}{m}\|Tx\| = \frac{1}{m}\|y\| ,$$

hence T^{-1}; $Y_1 = T(X) \to X$ is continuous.

(b) Assume conversely, that $T^{-1} \in \mathcal{L}(Y_1, X)$. Then, $\forall x \in X$, we get $x = T^{-1}(Tx)$ and

$$\|x\| \leq \left\|T^{-1}\right\| \|Tx\| , \quad \|Tx\| \geq \frac{1}{\left\|T^{-1}\right\|}\|x\| .$$

■

We shall now explain one of the fundamental results concerning families of linear continuous functions on Banach spaces.

Uniform boundedness theorem (U.B.T.) *Let X be a Banach space, Y a linear normed space, and let $\{T_\alpha\}_{\alpha \in \mathcal{A}}$ be a family (indexed by the set $\mathcal{A}$), of elements in $\mathcal{L}(X,Y)$. Assume that, $\forall x \in X$, the numerical set $\left\{\|T_\alpha x\|_Y, \ \alpha \in \mathcal{A}\right\}$ is bounded. Then the set of norm $\left\{\|T_\alpha\|_{\mathcal{L}(X,Y)}\right\}$ is also bounded.*

Proof Let us define sets in X: $B_n = \left\{x \in X, \ \|T_\alpha x\| \leq n \ \forall \alpha \in \mathcal{A}\right\}$. Each set B_n is closed and $X = \bigcup_1^\infty B_n$. As X is a complete metric space, from Baire's category theorem we derive that $\exists n_0 \in \mathbb{N}$ such that B_{n_0} contains a ball in X, say $\left\{x, \ \|x - x_0\| < r\right\}$. Then the proof continues in exactly the same way as in Proposition 2.1. ■

We now point out some consequences of the U.B.T. above.

(1) Let X and Y be two Banach spaces; $(T_n)_1^\infty$ a sequence of operators in $\mathcal{L}(X,Y)$ such that

$$\lim_{n\to\infty} T_n x \quad \text{exists (in } Y\text{)}, \ \forall x \in X . \tag{2.2}$$

Then the operator T, $X \to Y$, defined by: $Tx = \lim_{n\to\infty} T_n x$ is linear and continuous; the following inequality holds

$$\|T\|_{\mathcal{L}(X,Y)} \le \liminf \|T_n\|_{\mathcal{L}(X,Y)} . \tag{2.3}$$

(2) Let H be a Hilbert space and A a linear operator $H \to H$ satisfying the symmetry condition

$$(Ah,k) = (h,Ak) , \quad \forall h,k \in H . \tag{2.4}$$

Then A is continuous ($A \in \mathcal{L}(H,H) = \mathcal{L}(H)$).

Proof Consider the family of linear continuous functionals on H, $\{Fy\}_{\|y\|\le 1}$, which is defined by

$$Fy(x) = (x, Ay) , \quad \forall x \in H . \tag{2.5}$$

We note that

$$|Fy(x)| = |(Ax,y)| \le \|Ax\| \quad \text{for } \|y\| \le 1 .$$

The U.B.T. implies that $\exists C > 0$ such that

$$|Fy(x)| \le C\|x\| , \quad \forall x \in H, \ \|y\| \le 1 ,$$

that is $|(x,Ay)| \le C\|x\|$ if $x \in H$ and $\|y\| \le 1$. Take $x = Ay$ and obtain $\|Ay\|^2 \le C\|Ay\|$ which implies $\|Ay\| \le C$ for $\|y\| \le 1$. ■

(3) Let X be a Banach space, $[a,b]$ a compact interval in $\mathbb{R}$, and $A(t)$ a function $[a,b] \to \mathcal{L}(X)$. Assume that the function $t \in [a,b] \to A(t)x$ is continuous, $[a,b] \to X$, $\forall x \in X$. Then, $\exists C > 0$ such that $\|A(t)\|_{\mathcal{L}(X)} \le C \ \forall t \in [a,b]$.

In fact, from continuity, we find, $\forall x \in X$ a constant $C_x > 0$, such that $\|A(t)x\| \le C_x$ $\forall t \in [a,b]$. Then apply U.B.T.

3. Here we remember the definitions and properties of the adjoint operator in Hilbert spaces. In fact, the following holds.

Theorem 3.1 *(See [Young - Theorem 7.13, p.76). Let $A \in \mathcal{L}(E,F)$ where E,F are Hilbert spaces. There exists a unique operator $A^* \in \mathcal{L}(F,E)$ such that*

$$(Ax, y)_F = (x, A^* y)_E \,, \quad \forall x \in E, \ \forall y \in F \,. \tag{3.1}$$

A^* is called the *adjoint* of A. Note also that $A^{**} = A$ and $\|A^*\| = \|A\|$ (Young, Theorem 7.15, p.78).

We shall use adjoints in Hilbert space in order to continue the investigation of sesquilinear forms on Hilbert spaces (see Chapter I - §3).

Therefore, again let $\mathcal{H}$ be a Hilbert space over K and the function $(x,y) \to B(x,y)$ from $\mathcal{H} \times \mathcal{H}$ into K, satisfying (3.13)-(3.14)-(3.15) of Chapter I - §3.

As seen in I-Theorem 3.3, there exists one and only one operator $A \in \mathcal{L}(\mathcal{H})$, such that

$$B(x,y) = (Ax, y) \,, \quad \forall x, y \in \mathcal{H} \,.$$

Let us assume also the lower estimate

$$|B(x,x)| \geq \alpha \|x\|^2 \,, \quad \forall x \in \mathcal{H} \ \text{(with some } \alpha > 0) \,. \tag{3.2}$$

It follows that the *above operator A is injective and surjective.*

In fact, from (3.2) it follows that $\alpha\|x\|^2 \leq |(Ax,x)| \leq \|Ax\| \, \|x\|$; hence, if $x \neq \mathbf{0}$, then $\alpha\|x\| \leq \|Ax\|$ and this gives injectivity.

Next, let us note that, $\forall x \in \mathcal{H}$, $B(x,x) = (Ax,x) = (x, A^* x)$. Therefore we get (again from (3.2)) that $|(x, A^* x)| \geq \alpha \|x\|^2$, and $\alpha\|x\|^2 \leq \|x\| \, \|A^* x\|$. This shows that operator A^* too is injective.

We now proceed to prove surjectivity of the operator A. Let $\mathcal{H}_1 = A(\mathcal{H})$ be the range of the operator A. It is obviously a linear subspace of $\mathcal{H}$. It is, furthermore, a closed subspace of $\mathcal{H}$: in fact, let (x_n) be a sequence in $\mathcal{H}_1$ and $x_n \to x_0 \in \mathcal{H}$. We have $x_n = Ah_n$, $h_n \in \mathcal{H}$. As previously seen, we get $\alpha\|h_n - h_m\| \leq \|Ah_n - Ah_m\| = \|x_n - x_m\|$, $\forall n, m \in \mathbb{N}$.

Therefore, (h_n) is a Cauchy sequence in $\mathcal{H}$; hence, $\exists h_0 \in \mathcal{H}$, $h_n \to h_0$. Then $Ah_n \to Ah_0 = x_0$, and $x_0 \in \mathcal{H}_1$.

Let us assume that $\mathcal{H}_1$ is a strict subspace of $\mathcal{H}$. As seen previously, this will entail the existence of $h \in \mathcal{H}$, $h \neq \mathbf{0}$, and $h \perp \mathcal{H}_1$. Therefore we get $(h, Ax) = 0$, $\forall x \in \mathcal{H}$, that is $(A^* h, x) = \mathbf{0}$, $\forall x \in \mathcal{H}$ and $A^* h = \mathbf{0}$. This however contradicts the injectivity of the operator A^*. (Another proof: if $h \perp \mathcal{H}_1 \Rightarrow (Ah, h) = B(h,h) = 0$, then $\alpha \|h\|^2 \leq |B(h,h)| = 0$ and $h = \mathbf{0}$.)

Next, let us introduce a second Hilbert space $\mathcal{X}$; we shall assume that $\mathcal{H} \subset \mathcal{X}$ as a linear subspace, that the set of elements in $\mathcal{H}$ is dense in the space $\mathcal{X}$ and that, with some constant $C > 0$, we have

$$\|u\|_{\mathcal{X}} \leq C \|u\|_{\mathcal{H}}, \qquad \forall u \in \mathcal{H}. \tag{3.3}$$

(This last inequality means that the identical mapping of $\mathcal{H}$ into $\mathcal{X}$ is continuous!)

We next proceed to show how to associate this situation to the sesquilinear form $a(u,v)$ on $\mathcal{H}$[4] another linear operator Z, acting on a set $\mathcal{D}$ contained in $\mathcal{H}$ with range in $\mathcal{X}$. Let us then define

$\mathcal{D} = \{u \in \mathcal{H}$ such that the antilinear function $v \in \mathcal{H} \to a(u,v) \in K$ is continuous on $\mathcal{H}$ with the norm of $\mathcal{X}\}$.

Thus, in order to define $\mathcal{D}$ we consider the function $v \to a(u,v)$ from the metric subspace of $\mathcal{X}$ which is $\mathcal{H}$ to the field K.

We have accordingly that

$$\forall u \in \mathcal{D}, \quad \text{and} \quad \forall \varepsilon > 0, \ \exists \delta > 0 \quad \text{such that} \quad v \in \mathcal{H} \text{ and } \|v\|_{\mathcal{X}} < \delta \Rightarrow |a(u,v)| < \varepsilon .$$

Then, $\forall u \in \mathcal{D}$ and $\forall v \in \mathcal{H}$, $v \neq \mathbf{0}$, we have

$$\left\| \frac{v}{2\|v\|_{\mathcal{X}}} \cdot \delta \right\|_{\mathcal{X}} = \frac{\delta}{2} < \delta$$

hence

$$\left| a\left(u, \frac{v}{2\|v\|_{\mathcal{X}}} \cdot \delta \right) \right| < \varepsilon \quad \text{and} \quad |a(u,v)| < \frac{2}{\delta} \varepsilon \|v\|_{\mathcal{X}}, \qquad \forall v \in \mathcal{H} .$$

---

[4] Note the change of notation for sesquilinear forms!

It is seen that

$$\left|a(u,v_1)-a(u,v_2)\right|\le C\|v_1-v_2\|_{\mathcal{X}}\ ,\qquad \forall v_1,v_2\in\mathcal{H}\ .$$

Therefore, the uniformly continuous function $v\in\mathcal{H}\to a(u,v)\in K$ can be extended - uniquely - by continuity, to a continuous function $v\in\mathcal{X}\to a(u,v)\in K$, for all $u\in\mathcal{D}$ and now $|a(u,v)|\le C\|v\|_{\mathcal{X}}$, $\forall v\in\mathcal{X}$ (remember that $\mathcal{H}$ is dense in $\mathcal{X}$!). The *linear* function $v\in\mathcal{X}\to\overline{a(u,v)}$ is thus continuous on $\mathcal{X}$ and can be represented by

$$\overline{a}(u,v)=\left(v,Z_u\right)_{\mathcal{X}}\qquad \text{with a unique element } Z_u\in\mathcal{X},\ \forall v\in\mathcal{X}\ . \tag{3.4}$$

Thus we have an operator Z, $\mathcal{D}\to\mathcal{X}$, defined by: $u\in\mathcal{D}\to Z_u=Z_u$, such that

$$\overline{a}(u,v)=\left(v,Z_u\right)_{\mathcal{X}}\qquad \text{or}\qquad a(u,v)=\left(Z_u,v\right)_{\mathcal{X}}\ ,\qquad \forall u\in\mathcal{D},\ \forall v\in\mathcal{X}\ . \tag{3.5}$$

Actually, $\mathcal{D}$ is a linear subspace of $\mathcal{H}$ and Z is a linear operator, $\mathcal{D}\to\mathcal{X}$. (In fact, if $u_1,u_2\in\mathcal{D}$, $a(u_1+u_2,v)=a(u_1,v)+a(u_2,v)$ is continuous on $\mathcal{H}$ with the norm of $\mathcal{X}$. Also, $\left(Z(u_1+u_2),v\right)_{\mathcal{X}}=\left(Zu_1,v\right)_{\mathcal{X}}+\left(Zu_2,v\right)_{\mathcal{X}}$, $\forall v\in\mathcal{X}$ which implies that $Z(u_1+u_2)=Zu_1+Zu_2$.)

We can now state:

Theorem 3.2 *Let $a(u,v)$ be a sesquilinear form on $\mathcal{H}\times\mathcal{H}$ which satisfies the estimates:*

$$|a(u,v)|\le C\|u\|_{\mathcal{H}}\|v\|_{\mathcal{H}}\ ,\quad \forall u,v\in\mathcal{H}\qquad \textit{and}\qquad |a(u,u)|\ge\alpha\|u\|_{\mathcal{H}}^2\ ,\quad \forall u\in\mathcal{H}\ , \tag{3.6}$$

(where α is some positive number). Then, the equation $Zu=f$, with f given in $\mathcal{X}$, has a unique solution $u\in\mathcal{D}$.

Proof Let us first establish uniqueness: assume therefore that $Zu_1=Zu_2=f$ where $u_1,u_2\in\mathcal{D}$. Then $Z(u_1-u_2)=\mathbf{0}$ and $a(u_1-u_2,v)=\left(Z(u_1-u_2),v\right)_{\mathcal{X}}=0$, $\forall v\in\mathcal{X}$. Take $v=u_1-u_2$ and obtain $a(u_1-u_2,\ u_1-u_2)=0$; therefore, from (3.6) we infer that $\|u_1-u_2\|^2=0$, hence $u_1=u_2$.

Next, in order to prove existence, let us first make the following:

Remark If $u\in\mathcal{H}$ and $a(u,v)=(f,v)_{\mathcal{X}}$, $\forall v\in\mathcal{H}$, then $u\in\mathcal{D}$. In fact we have $|a(u,v)|=\left|(f,v)_{\mathcal{X}}\right|\le\|f\|_{\mathcal{X}}\|v\|_{\mathcal{X}}$; this means that the function $v\in\mathcal{H}\to a(u,v)$ is continuous on $\mathcal{H}$ with respect to the $\mathcal{X}$-norm.

Hence, we get $a(u,v)=(Zu,v)_{\mathcal{X}}=(f,v)_{\mathcal{X}}$, $\forall v\in\mathcal{H}$. As $\mathcal{H}$ is dense in $\mathcal{X}$, it follows also that $Zu=f$.

Therefore, in order to find a solution for the equation $Zu=f$ it will be sufficient to solve the equation $a(u,v)=(f,v)_{\mathcal{X}}$, $\forall v\in\mathcal{H}$, and find a solution $u\in\mathcal{H}$.

Consider the linear functional on $\mathcal{H}$: $v\in\mathcal{H}\to(v,f)_{\mathcal{X}}$. It is a continuous function on $\mathcal{H}$: $|(v,f)_{\mathcal{X}}|\le\|v\|_{\mathcal{X}}\|f\|_{\mathcal{X}}\le C\|v\|_{\mathcal{H}}\|f\|_{\mathcal{X}}$, $\forall v\in\mathcal{H}$ (in view of (3.3)). By the Riesz representation theorem in $\mathcal{H}$ we find a unique element in $\mathcal{H}$, denoted with Jf, in such a way that $(v,f)_{\mathcal{X}}=(v,Jf)_{\mathcal{H}}$, $\forall v\in\mathcal{H}$.

Thus, $(f,v)_{\mathcal{X}}=(Jf,v)_{\mathcal{H}}$, $\forall v\in\mathcal{H}$, where $Jf\in\mathcal{H}$, hence J is a mapping from $\mathcal{X}$ into $\mathcal{H}$.

On the other hand, we know the existence of the operator $A\in\mathcal{L}(\mathcal{H})$ such that $a(u,v)=(Au,v)_{\mathcal{H}}$ $\forall u,v\in\mathcal{H}$. The equation $a(u,v)=(f,v)_{\mathcal{X}}$, $\forall v\in\mathcal{H}$, thus becomes

$$(Au,v)_{\mathcal{H}}=(Jf,v)_{\mathcal{H}}\,,\qquad v\in\mathcal{H}\,. \tag{3.7}$$

Finally, we have seen that – under our assumptions – the operator A is surjective and injective $\mathcal{H}\to\mathcal{H}$. Therefore, given $Jf\in\mathcal{H}$, $\exists u\in\mathcal{H}$, such that $Au=Jf$. This solves (3.7), and equation $Zu=f$, too. ■

4. We next consider the 'adjoint' sesquilinear form and the operators associated to it. Thus, let $(u,v)\in\mathcal{H}\times\mathcal{H}\to a(u,v)\in K$ be a sesquilinear form. The adjoint form, denoted here by a^*, is defined by

$$a^*(u,v)=\overline{a}(v,u)\,,\qquad \mathcal{H}\times\mathcal{H}\to K\,. \tag{4.1}$$

It is again a sesquilinear form: for instance,

$$a^*(u_1+u_2,v)=\overline{a}(v,u_1+u_2)=\overline{a}(v,u_1)+\overline{a}(v,u_2)=a^*(u_1,v)+a^*(u_2,v)\ ;$$

$$a^*(\lambda u,v)=\overline{a}(v,\lambda u)=\overline{\overline{\lambda}a(v,u)}=\lambda\overline{a}(v,u)=\lambda a^*(u,v)\ ;$$

$$a^*(u,\lambda v)=\overline{a}(\lambda v,u)=\overline{\lambda a(v,u)}=\overline{\lambda}\overline{a}(v,u)=\overline{\lambda}a^*(u,v)\ .$$

Let us now assume that $a(u,v)$ is 'upper-bounded' and 'lower bounded' (as in (3.6) above). It follows that

$$\left|a^*(u,v)\right| = |\bar{a}(v,u)| = |a(v,u)| \le C\|u\|\,\|v\|\,, \qquad \forall u,v \in \mathcal{H}$$
$$\left|a^*(u,u)\right| = |a(u,u)| \ge \alpha\|u\|^2\,, \qquad \forall u \in \mathcal{H} \tag{4.2}$$

which means that $a^*(u,v)$ is also 'upper bounded' and 'lower bounded'. (If $a^*(u,v) = a(u,v),\ \forall u,v \in \mathcal{H}$, we have a symmetric form.)

Let us now denote by $\mathcal{A}$ the linear continuous operator $\mathcal{H} \to \mathcal{H}$ such that $a(u,v) = (\mathcal{A}u,v)_{\mathcal{H}},\ \forall u,v \in \mathcal{H}$ (I - Theorem 3.3). It follows then that

$$a^*(u,v) = \overline{a(v,u)} = \overline{(\mathcal{A}v,u)_{\mathcal{H}}} = (u,\mathcal{A}v)_{\mathcal{H}} = (\mathcal{A}^*u,v)_{\mathcal{H}}\,, \qquad \forall u,v \in \mathcal{H} \tag{4.3}$$

where $\mathcal{A}^*$ is the adjoint of $\mathcal{A}$ and $\mathcal{A}^* \in \mathcal{L}(\mathcal{H})$.

This means that the linear continuous operator, $\mathcal{H} \to \mathcal{H}$, associated to $a^*(u,v)$, is precisely $\mathcal{A}^*$ which is – as $\mathcal{A}$ – an injective and surjective operator.

Let us now prove:

Proposition 4.1 *If $a(u,v)$ is a sesquilinear form on $\mathcal{H} \times \mathcal{H}$, satisfying (3.6) above, the domain $\mathcal{D}$ of the operator Z associated to it is dense in $\mathcal{H}$.*

Proof Let $v \in \mathcal{H}$ such that $(u,v)_{\mathcal{H}} = 0,\ \forall u \in \mathcal{D}$. Use now the injective and surjective operator $\mathcal{A}^*$, $\mathcal{H} \to \mathcal{H}$; it has an inverse $(\mathcal{A}^*)^{-1}$, $\mathcal{H} \to \mathcal{H}$. Denote $w = (\mathcal{A}^*)^{-1}v,\ v = \mathcal{A}^*w$. Then $(u,v)_{\mathcal{H}} = (u,\mathcal{A}^*w)_{\mathcal{H}} = (\mathcal{A}u,w)_{\mathcal{H}} = a(u,w) = (Zu,w)_{\mathcal{X}} = 0$.

Now apply Theorem 3.2: $\forall f \in \mathcal{X},\ \exists u \in \mathcal{D}$ such that $Zu = f$. Therefore we get

$$(f,w)_{\mathcal{X}} = 0, \quad \forall f \in \mathcal{X}, \quad \text{hence} \quad w = \mathbf{0} \quad \text{and} \quad v = \mathbf{0}.$$

Thus, if $v \in \mathcal{H}$ and $v \perp \mathcal{D}$ then $v = \mathbf{0}$. This implies the density of $\mathcal{D}$ in $\mathcal{H}$ (otherwise the closure $\overline{\mathcal{D}}$ is a proper subspace of $\mathcal{H}$, and, as previously seen, there must exist an element $\ne \mathbf{0}$ in $\mathcal{H}$ which is orthogonal to $\overline{\mathcal{D}}$, hence to $\mathcal{D}$, which is not possible.) ■

Remark $\mathcal{D}$ is dense in $\mathcal{H}$ - in the $\mathcal{H}$-norm, hence in the $\mathcal{X}$-norm. Also, $\mathcal{H}$ is dense in $\mathcal{X}$ - in the $\mathcal{X}$-norm. Then we have (with closure in $\mathcal{X}$) $\overline{\mathcal{D}} = \mathcal{H}$, $\overline{\mathcal{H}} = \mathcal{X}$, hence $\overline{\mathcal{D}} = \overline{\overline{\mathcal{D}}} = \overline{\mathcal{H}} = \mathcal{X}$. Consequently $\mathcal{D}$ is dense in $\mathcal{X}$ too.

5. Up to now we have only defined and used the concepts of *adjoint* operators in Hilbert space for the case of linear *continuous* operators. It is important, however, as shown in many examples, to extend this concept in order to cover situations where the operator is not necessarily continuous.

Let E be a Hilbert space and let A be a linear operator whose domain of definition ($\mathcal{D}(A)$) is a linear subset of E, while the range of A is in E. Thus, we have $A\colon \mathcal{D}(A) \subset E \to E$.

We shall assume also that the domain $\mathcal{D}(A)$ is dense in E: $\overline{\mathcal{D}(A)} = E$.

Let us define a set $\mathcal{D}^*$ in E by the following:

$$\mathcal{D}^* = \{h \in E \text{ such that, } \exists k \in E, \text{ with property that } (Ax,h)_E = (x,k)_E, \quad \forall x \in \mathcal{D}(A)\} . \tag{5.1}$$

Remark The element k corresponding to $h \in \mathcal{D}^*$ is uniquely determined for if, say, k_1 and k_2 correspond to the same $h \in \mathcal{D}^*$, we would have $(Ax,h) = (x,k_1) = (x,k_2)$ $\forall x \in \mathcal{D}(A)$. Now, from the density of $\mathcal{D}(A)$ we derive $k_1 = k_2$.

Note The set $\mathcal{D}^*$ can also be characterized by:

$$\mathcal{D}^* = \{h \in E \text{ such that the function: } x \in \mathcal{D}(A) \to (Ax,h) \text{ is continuous}\} . \tag{5.2}$$

In fact, this function is linear, hence Lipschitz continuous; as $\mathcal{D}(A)$ is dense, we extend it to the whole of E and then apply the representation theorem, to get the k in (5.1).

Conversely, if $(Ax,h) = (x,k)$, $\forall x \in \mathcal{D}(A)$, we are in the situation of (5.2). We also see that $h = \mathbf{0}$ belongs to $\mathcal{D}^*$.

Thus, from the above considerations, an operator A^*, $\mathcal{D}^* \to E$ is well defined by

$$A^* h = k , \quad \forall h \in \mathcal{D}^* . \tag{5.3}$$

Proposition 5.1 *The set $\mathcal{D}^*$ is linear in E; the operator A^* is linear.*

Proof Let $h_1, h_2 \in \mathcal{D}^*$; thus, $(Ax,h_1) = (x,k_1)$ and $(Ax,h_2) = (x,k_2)$ $\forall x \in \mathcal{D}(A)$, hence, $(Ax,h_1 + h_2) = (x,k_1 + k_2)$, $\forall x \in \mathcal{D}(A)$, which shows that $h_1 + h_2 \in \mathcal{D}^*$.

Also, $A^*(h_1 + h_2) = k_1 + k_2 = A^* h_1 + A^* h_2$.

Next, if $\lambda \in K$, we get, for $h \in \mathcal{D}^*$: $(Ax, \lambda h) = \overline{\lambda}(Ax,h) = \overline{\lambda}(x,k) = (x, \lambda k)$; hence, $\lambda h \in \mathcal{D}^*$ and $A^*(\lambda h) = \lambda A^* h$. ■

We may note therefore that $\mathcal{D}^*$ is the domain of definition of the operator A^*: $\mathcal{D}(A^*) = \mathcal{D}^*$.

From (5.1) we derive: $(Ax,h) = (x, A^*h)\ \forall x \in \mathcal{D}(A),\ \forall h \in \mathcal{D}(A^*)$.

As indicated previously, operator A is not necessarily continuous. The operator A^* is always a *closed operator* (here this means the following: if (h_n) is a sequence in $\mathcal{D}(A^*)$, $h_n \to h_0 \in E$ and $A^*h_n \to k_0 \in E \Rightarrow h_0 \in \mathcal{D}(A^*)$ and $A^*h_0 = k_0$). It is readily established that we have $(Ax, h_n) = (x, A^*h_n)\ \forall x \in \mathcal{D}(A),\ n = 1,2,\ldots$. As $n \to \infty$ we obtain that $(Ax, h_0) = (x, k_0),\ \forall x \in \mathcal{D}(A)$, which means $h_0 \in \mathcal{D}(A^*),\ A^*h_0 = k_0$.

Note also:

Proposition 5.2 *Let* $A,\ \mathcal{D}(A) \subset E \to E$ *be a symmetric operator:* $(Ax,y) = (x,Ay)$ $\forall x,y \in \mathcal{D}(A)$. *Assume also that* $\mathcal{D}(A)$ *is dense in* E. *Then* $A \subset A^*$. *Conversely, if* $A - \mathcal{D}(A) \to E$, *with dense domain, is contained in its adjoint, then A is symmetric.*

Note Two linear operators L_1, L_2, with domains $\mathcal{D}(L_1),\ \mathcal{D}(L_2)$ in E are in the relation $L_1 \subset L_2$ iff: $\mathcal{D}(L_1) \subset \mathcal{D}(L_2)$ and $L_2 = L_1$ on $\mathcal{D}(L_1)$.

Proof of Proposition 5.2 If $(Ax,y) = (x,Ay),\ \forall x,y \in \mathcal{D}(A)$, we see that any $y \in \mathcal{D}(A)$ belongs to $\mathcal{D}^*$, and $A^*y = Ay$. Thus, $\mathcal{D}(A) \subset \mathcal{D}(A^*)$ and $A \subset A^*$. Conversely, let $A \subset A^*$. Then, $\mathcal{D}(A) \subset \mathcal{D}(A^*)$ and $(Ax,y) = (x, A^*y),\ \forall x \in \mathcal{D}(A),\ \forall y \in \mathcal{D}(A)$. Furthermore, $A^* = A$ on $\mathcal{D}(A)$, hence $(Ax,y) = (x,Ay),\ \forall x,y \in \mathcal{D}(A)$. ■

An important concept now arises: that of a (not necessarily continuous) self-adjoint linear operator.

Definition 5.1 *Let* $A,\ \mathcal{D}(A) \subset H \to H$ *be a linear operator with dense domain. Then, if* $A = A^*$ *we say that A is a self-adjoint operator.*

Thus, we infer that *self-adjoint operators are closed and symmetric operators.*

6. In this section we return to the study of sesquilinear forms, especially of the 'adjoint' sesquilinear forms (Section 4). Thus, we again have the couple of Hilbert spaces $\mathcal{H}$ and $\mathcal{K}$; the sesquilinear form $a(u,v)$ on $\mathcal{H} \times \mathcal{H}$ which is 'upper and lower bounded'; and the adjoint sesquilinear form $a^*(u,v) = \overline{a}(v,u)$. We have two operators associated to $a(u,v)$: $\mathcal{A} \in \mathcal{L}(\mathcal{H})$, and $Z,\ \mathcal{D} \subset \mathcal{H} \to \mathcal{K}$ such that $a(u,v) = (\mathcal{A}u, v)_{\mathcal{H}},\ \forall u,v \in \mathcal{H}$ and

$a(u,v)=(Zu,v)_{\mathcal{X}}$, $\forall u\in\mathcal{D}$, $v\in\mathcal{X}$ (in the second equality $a(u,v)$ is extended (for $u\in\mathcal{D}$), from $v\in\mathcal{H}$ to the whole $\mathcal{X}$).

As seen previously, one has also $a^*(u,v)=(\mathcal{A}^*u,v)$, $\forall u,v\in\mathcal{H}$, where $\mathcal{A}^*$, the adjoint of $\mathcal{A}$, is again in $\mathcal{L}(\mathcal{H})$. Now, a^* is also upper and lower bounded; define the set

$$\mathcal{D}^1=\{u\in\mathcal{H}\text{, such that the function } v\in\mathcal{H}\to a^*(u,v) \text{ is continuous in the } \mathcal{X}\text{-norm}\}\,.$$

This set is again dense in $\mathcal{H}$; furthermore, there is a linear operator Z^1, $\mathcal{D}^1\to\mathcal{X}$, defined by

$$a^*(u,v)=\left(Z^1u,v\right)_{\mathcal{X}},\quad \forall u\in\mathcal{D}_1,\ \forall v\in\mathcal{X}\,. \tag{6.1}$$

(Here again $a^*(u,v)$ is extended, for $u\in\mathcal{D}_1$, to the whole space $\mathcal{X}$.)

As in Theorem 3.2 the equation $Z^1u=f$ where f is given in $\mathcal{X}$, has a unique solution $u\in\mathcal{D}_1$.

Let us consider now $u\in\mathcal{D}$ and $v\in\mathcal{D}_1$. We obtain immediately

$$(Zu,v)_{\mathcal{X}}=a(u,v)\ ;\quad \left(Z^1v,u\right)_{\mathcal{X}}=a^*(v,u)=\overline{a}(u,v)=(v,Zu)_{\mathcal{X}}\ ,$$

that is

$$(Zu,v)_{\mathcal{X}}=\overline{\overline{a}}(u,v)=\left(u,Z^1v\right)_{\mathcal{X}},\quad \forall u\in\mathcal{D},\ \forall v\in\mathcal{D}^1\,. \tag{6.2}$$

If we now use the definition of the adjoint operator in Section 5 we see that, because of (6.2), $\mathcal{D}^1\subset\mathcal{D}(Z^*)$ and $(Zu,v)_{\mathcal{X}}=(u,Z^*v)_{\mathcal{X}}$, $\forall u\in\mathcal{D}$, $\forall v\in\mathcal{D}^1$. Thus, by now we have that $Z^1\subset Z^*$.

Consider now $v\in\mathcal{D}(Z^*)$. We then have

$$(Zu,v)_{\mathcal{X}}=(u,Z^*v)_{\mathcal{X}},\quad \forall u\in\mathcal{D},\ v\in\mathcal{D}(Z^*)\,. \tag{6.3}$$

On the other hand, let us now consider the equation $Z^1u=Z^*v$. It will have the (unique) solution $v_0\in\mathcal{D}^1=\mathcal{D}\left(Z^1\right)$. Thus, $Z^1v_0=Z^*v$. From (6.3)-(6.2) we infer now that

$$(Zu,v)_{\mathcal{X}}=\left(u,Z^1v_0\right)_{\mathcal{X}}=\left(Zu,v_0\right)_{\mathcal{X}}\quad \forall u\in\mathcal{D}(Z)\,. \tag{6.4}$$

As Z maps $\mathcal{D}$ onto $\mathcal{X}$, it follows that $v=v_0\in\mathcal{D}^1$.

Thus, $\mathcal{D}(Z^*)\subset\mathcal{D}\left(Z^1\right)$ and $Z^*=Z^1$.

The operators Z, associated with $a(u,v)$, and Z^1, associated with $a^*(u,v)$, are related by

$$Z^1 = Z^* . \tag{6.5}$$

Note the following as important:

Corollary *If the symmetric form $a(u,v)$ is lower and upper bounded the associated operator Z is self-adjoint.*

7. Concerning adjoints of discontinuous linear operators in Hilbert space we shall give a concrete example where the domain of the adjoint operator is not dense in the given space.

The space E (see Section 5) is now $L^2(\mathbb{R})$, the space of square integrable (in Lebesgue's sense) complex-valued functions defined on the real line; the scalar product is given by the relation

$$(f,g)_{L^2(\mathbb{R})} = \int_{\mathbb{R}} f(x)\overline{g}(x)\,\mathrm{d}x\,, \quad \forall f,g \in L^2(\mathbb{R})\,. \tag{7.1}$$

Let now D be the set of $f \in E$ such that $\int_{\mathbb{R}} |f(x)|\,\mathrm{d}x < +\infty$; thus $D = L^2(\mathbb{R}) \cap L^1(\mathbb{R})$, which is a dense set in E (it contains, in particular, continuous functions with compact support in $\mathbb{R}$).

Let us also fix a function $\varphi_0(\cdot) \in L^2(\mathbb{R})$, $\varphi_0 \neq 0$, and then let us define an operator T, $\varphi \in D \to T\varphi \in E$, by the formula

$$T\varphi = \left(\int_{\mathbb{R}} \varphi(x)\,\mathrm{d}x \right) \cdot \varphi_0\,, \quad \forall \varphi \in D\,. \tag{7.2}$$

We see that it is a linear operator in E, with dense domain.

Assuming that $\psi(\cdot) \in \mathcal{D}(T^*)$ we must have, according to (5.1), the equality

$$(T\varphi, \psi)_{L^2} = (\varphi, T^*\psi)_{L^2}\,, \quad \forall \varphi \in D\,. \tag{7.3}$$

The left-hand side in (7.3) can be written as:

$$\int_{\mathbb{R}} (T\varphi)(x)\overline{\psi}(x)\,\mathrm{d}x = \int_{\mathbb{R}}\left(\int_{\mathbb{R}} \varphi(t)\,\mathrm{d}t\right)\cdot\varphi_0(x)\cdot\overline{\psi}(x)\,\mathrm{d}x = \int_{\mathbb{R}} \varphi(x)\left(\int_{\mathbb{R}} \varphi_0(t)\overline{\psi}(t)\,\mathrm{d}t\right)\mathrm{d}x \quad (7.4)$$

(an absolutely convergent integral: $\varphi_0,\ \psi \in L^2 \Rightarrow \varphi_0 \cdot \overline{\psi} \in L^1$, and φ belongs to L^1 by the assumptions on D).

As (7.4) holds $\forall \varphi \in C_0(\mathbb{R})$ we get from (7.3):

$$\int_{\mathbb{R}} \varphi(x)\left(\int_{\mathbb{R}} \varphi_0(t)\overline{\psi}(t)\,\mathrm{d}t\right)\mathrm{d}x = \int_{\mathbb{R}} \varphi(x)\overline{(T^*\psi)}(x)\,\mathrm{d}x\,, \qquad \forall \varphi \in C_0(\mathbb{R}) \qquad (7.5)$$

and *this* in turn implies that

$$\overline{(T^*\psi)}(x) = \int_{\mathbb{R}} \varphi_0(t)\overline{\psi}(t)\,\mathrm{d}t\,, \qquad \text{a}\cdot\text{e in } \mathbb{R}$$

which of course is impossible (the right-hand side is constant on $\mathbb{R}$; the left-hand side should be in $L^2(\mathbb{R})$) unless the right-hand side is 0.

We have therefore:

$$\psi \in \mathcal{D}(T^*) \Rightarrow \psi \perp \varphi_0 \quad \text{and} \quad T^*\psi = 0\,,$$

and also

$$\psi \perp \{\lambda\varphi_0\}_{\lambda\in K} \quad \text{and} \quad T^*\psi = 0\,; \quad \mathcal{D}(T^*) \subset \{\lambda\varphi_0\}^{\perp}\,.$$

Conversely, let $\psi \in \{\lambda\varphi_0\}^{\perp}$; then $(\varphi_0, \psi) = 0$ and, $\forall \varphi \in D$, $(T\varphi, \psi) = 0$ (using (7.4)). This shows that $\psi \in \mathcal{D}(T^*)$ and $T^*\psi = 0$.

In this way we see that the domain of T^* is the orthogonal space to $\{\lambda\varphi_0\}$, which is a *closed* strict subspace of E; *it cannot be dense in* E!

8. Here we shall resume the general discussion of adjoint operators in Section 5.

First, let us note the following.

Proposition 8.1 *In the Hilbert space H let us consider linear operators A and B, where $A \subset B$ and $\mathcal{D}(A)$ is dense in H. Then $B^* \subset A^*$.*

Proof Both $\mathcal{D}(A)$ and $\mathcal{D}(B)$, which contains $\mathcal{D}(A)$, are dense in H. Therefore, the adjoint operators A^* and B^* are both well defined.

Let now $h \in \mathcal{D}(B^*)$ and $k \in \mathcal{D}(A)$. We have

$$(Ak, h) = (Bk, h) = (k, B^* h) \tag{8.1}$$

as $A \subset B$. This equality shows that $h \in \mathcal{D}(A^*)$ and $A^* h = B^* h$. Therefore $\mathcal{D}(B^*) \subset \mathcal{D}(A^*)$ and $A^* = B^*$ on $\mathcal{D}(B^*)$. ■

Of importance is the following:

Theorem 8.1 *Let A be symmetric operator with dense domain $\mathcal{D}(A)$ in the Hilbert space H. Then the domain of A^* is also dense in H and $A \subset (A^*)^*$.*

Proof As seen in Proposition 5.2, we have $A \subset A^*$, hence $\mathcal{D}(A^*) \supset \mathcal{D}(A)$ and $\overline{\mathcal{D}(A^*)} = H$. This shows the existence of the 'double adjoint' operator $(A^*)^* = A^{**}$.

Let now $h \in \mathcal{D}(A)$ and $k \in \mathcal{D}(A^*)$. It results that

$$(Ah, k) = (h, A^* k) \quad \text{or else} \quad (A^* k, h) = (k, Ah) \tag{8.2}$$

and this is true $\forall k \in \mathcal{D}(A^*)$. It follows then - according to the definition of adjoints - that $h \in \mathcal{D}(A^{**})$ and $A^{**} h = Ah$. This has been established $\forall h \in \mathcal{D}(A)$, whence we may infer that $A^{**} \supset A$. ■

Remarks

(1) From $A \subset A^*$ and Proposition 8.1 we may derive $A^{**} \subset A^*$.

(2) The domain $\mathcal{D}(A^{**})$ contains $\mathcal{D}(A)$; hence, it is dense in H. Also, A^{**} is also symmetric: from $A^{**} \subset A^*$ it results - again by Proposition 8.1 - that $A^{**} \subset (A^{**})^*$; then apply Proposition 5.2.

Let now T_1, T_2 be linear operators in the Hilbert space E with domains $\mathcal{D}(T_1)$ and $\mathcal{D}(T_2)$. The sum of these operators, $T = T_1 + T_2$ will be defined on the domain $\mathcal{D}(T) = \mathcal{D}(T_1) \cap \mathcal{D}(T_2)$ by

$$Tx = T_1 x + T_2 x, \quad \forall x \in \mathcal{D}(T). \tag{8.3}$$

We can establish the following:

Proposition 8.2 *Let T_1, T_2 be linear operators in H with dense domains $\mathcal{D}(T_1)$ and $\mathcal{D}(T_2)$ and assume also that $\mathcal{D}(T_1)\cap\mathcal{D}(T_2)$ is dense too. Then*

$$(T_1+T_2)^* \supset T_1^* + T_2^* \ . \tag{8.4}$$

Proof Let $x \in \mathcal{D}(T_1^* + T_2^*) = \mathcal{D}(T_1^*)\cap\mathcal{D}(T_2^*)$. Therefore, $\forall y \in \mathcal{D}(T_1)\cap\mathcal{D}(T_2)$ we have:

$$(T_1 y, x) = (y, T_1^* x) \ , \qquad (T_2 y, x) = (y, T_2^* x)$$

and by addition

$$((T_1+T_2)y, x) = (y, T_1^* x + T_2^* x) \ , \qquad \forall y \in \mathcal{D}(T_1)\cap\mathcal{D}(T_2) = \mathcal{D}(T_1+T_2) \ .$$

This shows that $x \in \mathcal{D}((T_1+T_2)^*)$ and that $(T_1+T_2)^* x = T_1^* x + T_2^* x$; whence $T_1^* + T_2^* \subset (T_1+T_2)^*$. ■

We terminate this section with:

Proposition 8.3 *Let T be a linear operator with dense domain $\mathcal{D}(T)$ in the Hilbert space H, and let us assume that $\mathcal{D}(T^*)$ is also dense in H. Then $T \subset (T^*)^*$.*

Note One should compare with Theorem 8.1 above! The operator T^{**} appears as a linear closed extension of *T*.

Proof Let $x \in \mathcal{D}(T)$, $y \in \mathcal{D}(T^*)$. We have

$$(Tx, y) = (x, T^* y) \quad \text{or else} \quad (T^* y, x) = (y, Tx) \ , \qquad \forall y \in \mathcal{D}(T^*) \ .$$

Thus, $x \in \mathcal{D}(T^{**})$ and $T^{**} x = Tx$. ■

9. Let us consider again (in this section) linear operators between two Banach spaces, *E*,*F* over the same field. More precisely, let *D* be a vector subspace of *E* and then *A* be a linear mapping, $D \to F$. Thus *D* is the domain of *A*, $\mathcal{D}(A) = D$.

The graph of the operator A (denoted by $G(A)$) is the following subset of the cartesian product $E \times F$:

$$G(A) = \{u, Au\}_{u \in D} \ . \tag{9.1}$$

The range of the operator A is the set $R(A) = \{Au\}_{u \in D}$. It is a linear subspace of F, as is readily seen. The kernel of A, Ker A, is the set $\{u \in D,\ Au = \mathbf{0}\}$, again, a linear subject of D.

Let us give the following:

Definition 9.1 *The linear operator A, $\mathcal{D}(A) \subset E \to F$, is said to be closed if $G(A)$ is a closed subset of $E \times F$.*

(Note that we consider here the Banach space $E \times F$ where the norm of a couple (e, f) is defined as: $\|(e, f)\| = \|e\|_E + \|f\|_F$.)

The connection with the concept of a closed operator encountered in Proposition 5.1 (the adjoint operator in Hilbert spaces is always closed) is easily made in the following:

Proposition 9.1 *The linear operator A, $\mathcal{D}(A) \subset E \to F$, is a closed operator iff $\forall$ sequences $(u_n)_1^\infty$ in $\mathcal{D}(A)$, such that $u_n \to u_0$ in E, while $Au_n \to v_0$ in F, it results that $u_0 \in \mathcal{D}(A)$ and $Au_0 = v_0$.*

Proof

(i) Let us assume that $G(A)$ is closed in $E \times F$. This means that, if $(u_n,\ Au_n)_1^\infty$ is a sequence in $G(A)$ which converges in $E \times F$ towards (u_0, v_0), then $(u_0, v_0) \in G(A)$ too.

Now, (u_n, Au_n) is convergent to (u_0, v_0) iff $u_n \to u_0$ in E and $Au_n \to v_0$ in F. Here $u_n \in \mathcal{D}(A)$, $\forall n \in \mathbb{N}$. The conclusion $(u_0, v_0) \in G(A)$ means that $u_0 \in \mathcal{D}(A)$ and $v_0 = Au_0$.

(ii) Let us assume the property after 'iff', and then take a sequence (u_n, v_n) in $G(A)$ such that $(u_n, v_n) \to (u_0, v_0)$.

This means that $u_n \in \mathcal{D}(A)$, $\forall n \in \mathbb{N}$, $v_n = Au_n$, $u_n \to u_0$, $Au_n \to v_0$.

By the assumption, $u_0 \in \mathcal{D}(A)$ and $v_0 = Au_0$, that is $(u_0, v_0) \in G(A)$, too. ■

Remark 1 If A is a (linear) closed operator then Ker A is a closed subset of E.

In fact, let $u_n \in \text{Ker } A$, $\forall n \in \mathbb{N}$, and $u_n \to u_0 \in E$. We have also $Au_n = \mathbf{0}$ and then (as A is closed) it results that $u_0 \in \mathcal{D}(A)$ and $Au_0 = \mathbf{0}$, which means that $u_0 \in \text{Ker } A$.

Remark 2 Let D be a closed linear subspace of E and A be a linear continuous operator, $D \to F$. Then A is a closed operator. In fact, let $(u_n)_1^\infty$ a sequence in D, such that $u_n \to u_0$ and $Au_n \to v_0$. As D is closed in E it follows that $u_0 \in D$. As A is continuous, it follows that $Au_n \to Au_0$. Thus $v_0 = Au_0$.

Example (A closed linear operator which is not continuous)

Let $E = F = C[0,1]$, the Banach space of all continuous functions, $[0,1] \to \mathbb{R}$ with the norm

$$\|\varphi(\cdot)\| = \sup_{0 \le t \le 1}, \quad |\varphi(t)| \ \forall \varphi \in C[0,1] .$$

Next, let D be the linear subspace of E composed of continuously differentiable functions $[0,1] \to \mathbb{R}$.

Let A be the operator, $D \to E$, defined by ordinary differentiation:

$$(A\varphi)(t) = \varphi'(t) , \qquad \forall t \in [0,1], \ \forall \varphi \in D . \tag{9.2}$$

Thus A is (obviously) a linear operator.

Note that A is a discontinuous operator: let $\varphi_n(t) = t^n$, $0 \le t \le 1$. This is a sequence of elements in E, such that $\|\varphi_n(\cdot)\| = \sup_{0 \le t \le 1} t^n = 1$.

Also, $(A\varphi_n)(t) = \varphi_n'(t) = nt^{n-1}$, hence $\sup_{0 \le t \le 1} |(A\varphi_n)(t)| = \|A\varphi_n\|_E = \sup_{0 \le t \le 1} \left(nt^{n-1}\right) = n$.

Thus the operator A is indeed discontinuous.

It is however a closed operator: in fact, let $(\varphi_n(\cdot))$ be a sequence in D, such that $\varphi_n(\cdot) \to \varphi_0(\cdot)$ in $C[0,1]$, while $A\varphi_n = \varphi_n' \to \psi_0$ in $C[0,1]$, too.

Now, we see that

$$\varphi_n(t) = \int_0^t \varphi_n'(s)\, \mathrm{d}s + \varphi_n(0) , \qquad n = 1,2,\ldots, \ t \in [0,1] . \tag{9.3}$$

From the uniform convergence on [0,1] of $\varphi_n'(\cdot)$ towards $\psi_0(\cdot)$, we get, from (9.3), the relation

$$\varphi_0(t) = \int_0^t \psi_0(s)\,\mathrm{d}s + \varphi_0(0) . \tag{9.4}$$

Thus, φ_0' exists and equals ψ_0, which means $\varphi_0 \in D$ and $A\varphi_0 = \psi_0$. A is closed. ■

A slightly more general class of linear operators is composed of so-called 'closable' operators.

Definition 9.2 *The linear operator T, $\mathcal{D}(T) \subset E \to F$, is said to be closable if it has a closed extension (an operator $\tilde{T}$, $\mathcal{D}(\tilde{T}) \subset E \to F$, which is closed, and such that $T \subset \tilde{T}$).*

In relation to closable operators the following holds.

Theorem 9.1 *Let T be a linear operator, $\mathcal{D}(T) \subset E \to F$, where E,F are Banach spaces. Then T is closable if and only if $\forall$ sequences (x_n) in $\mathcal{D}(T)$, such that $x_n \to \mathbf{0}$ and $Tx_n \to y$, it results that $y = \mathbf{0}$ (thus $(\mathbf{0}, y) \in \overline{G(T)} \Rightarrow y = \mathbf{0}$).*

Proof

(i) Let us assume that T is closable, and let $\tilde{T}$ be a closed operator which is an extension of T. It results that (x_n) is a sequence in $\mathcal{D}(\tilde{T})$, such that $x_n \to \mathbf{0}$ and $\tilde{T}x_n \to y$ and this implies that $y = \tilde{T}\mathbf{0} = \mathbf{0}$.

(ii) Let us assume the supposition of Theorem 9.1. We shall now define an operator S in the following way:

$$\mathcal{D}(S) = \{x \in E \text{ such that } \exists \text{ a sequence } (x_n) \text{ in } \mathcal{D}(T), \text{ where } x_n \to x \text{ while } Tx_n \to y\} \tag{9.5}$$

and then

$$Sx = y . \tag{9.6}$$

We have to establish several facts concerning this operator S.

(a) S is a well-defined, single-valued operator; in fact, let (x_n') be another sequence in $\mathcal{D}(T)$ where $x_n' \to x$, while $Tx_n' \to y'$. Then $(x_n - x_n') \to \mathbf{0}$ and $T(x_n - x_n') \to y - y'$. Therefore, $y - y' = \mathbf{0}$, $y = y'$.

(b) S is a linear operator: let $x_1, x_2 \in \mathcal{D}(S)$; hence $\exists$ a sequence $(x_{1,n})$ in $\mathcal{D}(T)$, where $x_{1,n} \to x_1$ while $Tx_{1,n} \to y_1 = Sx_1$; also $\exists$ a sequence $(x_{2,n})$ in $\mathcal{D}(T)$ where $x_{2,n} \to x_2$ while $Tx_{2,n} \to y_2 = Sx_2$. Now, we see that the sequence $(x_{1,n} + x_{2,n})$ in $\mathcal{D}(T)$ converges to $x_1 + x_2$ while $T(x_{1,n} + x_{2,n}) = T(x_{1,n}) + T(x_{2,n})$ converges to $y_1 + y_2$; thus $x_1 + x_2 \in \mathcal{D}(S)$ and $S(x_1 + x_2) = y_1 + y_2 = Sx_1 + Sx_2$.

In a similar way we see that

$$S(\lambda x) = \lambda S(x) \quad \forall x \in \mathcal{D}(S),\ \forall \lambda \in K .$$

(c) S is an extension of T; in fact, if $x \in \mathcal{D}(T)$, then, taking sequence (x_n) where $x_n = x$, $\forall n \in \mathbb{N}$, we see that $Tx_n = Tx \to Tx$. Hence, $x \in \mathcal{D}(S)$ and $Sx = Tx$.

(d) S is a closed operator: in fact, let (w_n) be a sequence in $\mathcal{D}(S)$ where $w_n \to w$ while $Sw_n \to u$. It follows that there exist sequences $(x_{n,p})_{p=1}^{\infty}$ in $\mathcal{D}(T)$, such that $x_{n,p} \to w_n$ (as $p \to \infty$) while $Tx_{n,p} \to Sw_n$.

Therefore, $\forall \varepsilon > 0$ we get: $\|x_{n,p} - w_n\| < \varepsilon$ and $\|Tx_{n,p} - Sw_n\| < \varepsilon$ provided $p \geq \bar{p}(\varepsilon, n)$, $\forall n = 1,2,\ldots$.

Let us take now $\varepsilon = 1/n$ and $p_n = \bar{p}(1/n, n)$. We obtain:

$$\|x_{n,p_n} - w_n\| < \frac{1}{n} \quad \text{and} \quad \|Tx_{n,p_n} - Sw_n\| < \frac{1}{n}, \qquad \forall n \in \mathbb{N} .$$

Here we note that $x_{n,p_n} \in \mathcal{D}(T)$, $\forall n \in \mathbb{N}$, and furthermore:

$$\|x_{n,p_n} - w\| \leq \|x_{n,p_n} - w_n\| + \|w_n - w\| < \frac{1}{n} + \|w_n - w\|$$

which $\to 0$ as $n \to \infty$. We also see that

$$\|Tx_{n,p_n} - u\| \leq \|Tx_{n,p_n} - Sw_n\| + \|Sw_n - u\| < \frac{1}{n} + \|Sw_n - u\|$$

which $\to 0$ as $n \to \infty$.

According to the definition of $\mathcal{D}(S)$ and S, this means that $w \in \mathcal{D}(S)$, while $Sw = u$. Hence, S is a closed extension of T. ■

Remark 1 We can see that the graph $G(S)$ is the closure of $G(T)$: $G(S) = \overline{G(T)}$. In fact, $\overline{G(T)}$ is composed of pairs (x, y) in $E \times F$ such that, for some sequence (x_n, Tx_n) in $G(T)$, it is true that $x_n \to x$, $Tx_n \to y$.

This means exactly that $\overline{G(T)} = G(S)$.

Remark 2 The previous constructed operator S is the minimal closed extension of the operator T in the sense that $\tilde{T}$ closed linear and $\tilde{T} \supset T \Rightarrow S \subset \tilde{T}$. In fact, let $x \in \mathcal{D}(S)$; then, $\exists$ sequence (x_n) in $\mathcal{D}(T)$ hence in $\mathcal{D}(\tilde{T})$ where $x_n \to x$ while $Tx_n = \tilde{T}x_n \to y = Sx$. This shows - as $\tilde{T}$ is closed - that $x \in \mathcal{D}(\tilde{T})$ and $\tilde{T}x = Sx$. Hence $S \subset \tilde{T}$.

In view of the previous discussions, we can say that S is the closure of T, and write $S = \overline{T}$. Then, Remark 1 now reads $G(\overline{T}) = \overline{G(T)}$.

We shall now give the following:

Example (a closable operator in $L^2(\Omega)$). Let Ω be an open subset of $\mathbb{R}^n$; take $E = F = L^2(\Omega)$.

Let

$$D = \left\{ f \in L^2(\Omega) \cap C^1(\Omega), \text{ such that } \frac{\partial f}{\partial x_j} \in L^2(\Omega) \right\}.$$

Define T_j, $D \to L^2(\Omega)$ by

$$T_j f = \frac{\partial f}{\partial x_j}.$$

Then T_j is closable; in fact, let (f_p) be a sequence in D, such that $f_p \to \mathbf{0}$ in $L^2(\Omega)$, while $\partial f_p / \partial x_j \to g$ in $L^2(\Omega)$.

Take arbitrary $\varphi \in C_0^1(\Omega)$ (once-continuously differentiable functions in Ω with compact support $K_\varphi \subset \Omega$). One obtains, using integration by parts:

$$\int_\Omega \left(\frac{\partial f_p}{\partial x_j} \right)(x) \varphi(x)\, \mathrm{d}x = -\int_\Omega f_p(x) \frac{\partial \varphi}{\partial x_j}\, \mathrm{d}x .$$

As $p \to \infty$ we readily obtain

$$\int_{\Omega} g(x)\varphi(x)\,\mathrm{d}x = 0 \ .$$

As this holds $\forall \varphi \in C_0^1(\Omega)$ we deduce that $g(x) = 0$ a·e in Ω, hence $g = \mathbf{0}$ in $L^2(\Omega)$.

10. We shall now briefly discuss a fundamental result about closed operators, the so-called:

Closed graph theorem *Let T be a closed operator, $E \to F$, where both E and F are Banach spaces. (Thus $\mathcal{D}(T) = E$.)*

Then T is a continuous operator.

Proof The graph of T, $G(T) = \{x, Tx\}_{x \in E}$, is a closed linear subspace of the (Banach) space $E \times F$. Hence $G(T)$ itself will be a Banach space.

Consider next the mapping U from $G(T)$ into E whose definition is

$$U\{x, Tx\} = x \ . \tag{10.1}$$

It is readily seen that U is a linear, injective, surjective, and continuous mapping:

$$\begin{aligned} U\big(\{x_1, Tx_1\} \ + \ \{x_2, Tx_2\}\big) &= U\big(\{x_1 + x_2,\ T(x_1 + x_2)\}\big) \\ &= x_1 + x_2 \\ &= U\{x_1, Tx_1\} + U\{x_2, Tx_2\} \ . \end{aligned}$$

If $U\{x, Tx\} = \mathbf{0}$, this means that $x = \mathbf{0}$, and $\{x, Tx\} = \{\mathbf{0}, \mathbf{0}\}$. For every $x_0 \in E$, $U\{x_0, Tx_0\} = x_0$.

Finally

$$\|U\{x, Tx\}\|_E = \|x\|_E \le \|x\|_E + \|Tx\|_F = \|\{x, Tx\}\|_{E \times F}$$

which shows that U is a continuous mapping.

Consider also the transformation V, from $G(T)$ into $R(T)$, the range of T in F, defined by

$$V\{x,Tx\}=Tx\,,\qquad \forall x\in E\,. \tag{10.2}$$

Then

$$\|V\{x,Tx\}\|_F=\|Tx\|_F\le\|\{x,Tx\}\|_{E\times F}$$

whence V is also a linear continuous mapping.

Turning back to the injective mapping U, it will have an inverse mapping U^{-1}, $E\to G(T)$ (remember that U is surjective). Then

$$U^{-1}x=\{x,Tx\}\,,\qquad \forall x\in E\,. \tag{10.3}$$

These formulas show the representation

$$Tx=VU^{-1}x\,,\qquad \forall x\in E\,. \tag{10.4}$$

At this stage we use the following 'high-powered' theorem (see for instance, A. Friedman, *Found. Mod. Anal.*, Theorem 4.6.2, p. 143).

Let X,Y be Banach spaces and L be an injective and surjective continuous linear mapping, $X\to Y$. Then L^{-1} is also a continuous linear mapping.

In our present situation $X=G(T)$, $Y=E$, $L=U$. Hence, U^{-1} also appears to be a continuous mapping, so that, from (10.4), we derive that $T=VU^{-1}$ is continuous, $E\to F$. ∎

The present section terminates with:

Proposition 10.1 *Let E,F be Banach spaces and T, $\mathcal{D}(T)\subset E\to F$ be a closed linear operator, such that T is injective. Then, the inverse operator T^{-1}, $R(T)\subset F\to D(T)\subset E$ is also a closed operator.*

Proof We have $G(T)=\{x,Tx\}_{x\in\mathcal{D}(T)}$; $G\left(T^{-1}\right)=\left\{y,T^{-1}y\right\}_{y\in R(T)}$.

Note that, if $T^{-1}y=z\in\mathcal{D}(T)$, then $y=Tz$. Thus we can also write

$$G\left(T^{-1}\right)=\{Tz,z\}_{z\in\mathcal{D}(T)}\,.$$

We see that $G(T)$ is a set in $E \times F$, while $G(T^{-1})$ is a set in $F \times E$.

Note that the mapping P, $\{x, y\} \to \{y, x\}$, from $E \times F$ onto $F \times E$ is a homeomorphism (continuous with continuous inverse); this implies that P maps closed sets in $E \times F$ into closed sets in $F \times E$; as $G(T)$ is closed in $E \times F$ and $G(T^{-1}) = PG(T)$, we see that $G(T^{-1})$ is closed in $F \times E$. ■

11. In the present section we again consider linear operators in Hilbert spaces and their adjoints as in previous sections.

Our first result here appears as:

Theorem 11.1 *Let T be a linear operator with dense domain in the Hilbert space H. Assume furthermore that T is closable. Then, the domain of the adjoint operator:* $\mathcal{D}(T^*)$ *is also dense in H.*

In the proof below we make use of some simple geometrical ideas in cartesian products of Hilbert spaces. First

Definition 11.1 Let $\mathcal{H}$ be a Hilbert space and $\mathcal{H} \times \mathcal{H}$ its square - the cartesian product with itself. The operator $\mathcal{V}$ on $\mathcal{H} \times \mathcal{H}$ is defined by

$$\mathcal{V}\{x, y\} = \{-y, x\}, \quad \forall x, y \in \mathcal{H} . \tag{11.1}$$

Note that the scalar product on $\mathcal{H} \times \mathcal{H}$ is naturally defined by $(\{x_1, y_1\}, \{x_2, y_2\})_{\mathcal{H} \times \mathcal{H}} = (x_1, x_2)_{\mathcal{H}} + (y_1, y_2)_{\mathcal{H}}$, so that

$$\|\{x_1, y_1\}\|^2 = (\{x_1, y_1\}, \{x_1, y_1\}) = \|x_1\|^2 + \|y_1\|^2 \tag{11.2}$$

gives the (Hilbertian induced) norm on $\mathcal{H} \times \mathcal{H}$.

We readily see that $\mathcal{V}$ maps $\mathcal{H} \times \mathcal{H}$ onto itself; $\mathcal{V}^2\{x, y\} = -\mathcal{I}$ (where $\mathcal{I}$ is the identity operator on $\mathcal{H} \times \mathcal{H}$). $\mathcal{V}$ is isometric

$$\|\mathcal{V}\{x, y\}\| = \|\{x, y\}\| . \tag{11.3}$$

Let us now place ourselves in the Hilbert space H and let T: $\mathcal{D}(T) \subset H \to H$ be the linear operator in Theorem 11.1.

Thus, $G(T)$ is a set in $H \times H$. We see that $G(T) = \{x, Tx\}_{x \in \mathcal{D}(T)}$ so that

$$\mathcal{V}(G(T)) = \{-Tx, x\}_{x \in \mathcal{D}(T)} \,. \tag{11.4}$$

We shall prove the following:

Lemma 11.1 *Let T be a linear operator with dense domain in the Hilbert space H, and T^* its (Hilbertian) adjoint operator. Then*

$$\left[\mathcal{V}(G_T)\right]^{\perp} = G_{T^*} \tag{11.5}$$

holds true (the orthogonality sign $^{\perp}$ is taken in $H \times H$).

In fact:

(a) Let $x \in \mathcal{D}(T^*)$; then $\{x, T^* x\} \perp \{-Ty, y\}$, $\forall y \in \mathcal{D}(T)$. (The corresponding scalar product is

$$(x, -Ty)_H + (T^* x, y)_H = -\overline{(Ty, x) + (y, T^* x)}$$

which equals 0 $\forall y \in \mathcal{D}(T)$ and $x \in \mathcal{D}(T^*)$ from the very definition of T^*.) Therefore we can say that: $G_{T^*} \subset \left[\mathcal{V}(G_T)\right]^{\perp}$.

(b) Let $\{x, y\} \in H \times H$ such that $\{x, y\} \perp \mathcal{V}(G_T)$, that is

$$\{x, y\} \perp \{-Tu, u\} \,, \qquad \forall u \in \mathcal{D}(T) \,.$$

Then we get $-(x, Tu)_H + (y, u)_H = 0$, $\forall u \in \mathcal{D}(T)$, that is $(Tu, x) = (u, y)$, $\forall u \in \mathcal{D}(T)$.

Again, the definition of T^* implies that $x \in \mathcal{D}(T^*)$ and $T^* x = y$. Thus $\{x, y\} = \{x, T^* x\}$ with $x \in \mathcal{D}(T^*)$, that is $\{x, y\} \in G_{T^*}$.

Consequently we obtain the reverse inclusion

$$\left[\mathcal{V}(G_T)\right]^{\perp} \subset G_{T^*} \,.$$

This proves (11.5). ■

Next we need a similar result, expressed here as:

Lemma 11.2 *If T is a linear operator with dense domain in the Hilbert space H and T^* its adjoint operator we have, in $H \times H$*

$$(G_T)^{\perp} = \mathcal{V}\left(G_{T^*}\right). \tag{11.6}$$

The result is proved in two steps, as in the previous lemma.

(a) Let $\{x,y\} \in (G_T)^{\perp}$, which means that $\{x,y\} \perp \{u,Tu\}$, $\forall u \in \mathcal{D}(T)$, and this translates as $(x,u)+(y,Tu)=0$, $\forall u \in \mathcal{D}(T)$ or also $(Tu,y)=-(u,x)$, $\forall u \in \mathcal{D}(T)$.

As usual, this last equality entails: $y \in \mathcal{D}(T^*)$, $T^* y = -x$; thus, $\{x,y\} = \{-T^* y, y\}$ which belongs to $\mathcal{V}\left(G_{T^*}\right)$.

(b) Let $\{x,y\} \in \mathcal{V}\left(G_{T^*}\right)$; therefore $\{x,y\} = \{-T^* u, u\}$ for some $u \in \mathcal{D}(T^*)$.

Take now any element $\{z,Tz\}$ in G_T. It follows accordingly that $(\{-T^* u,u\}, \{z,Tz\})_{H\times H} = (-T^* u,z)_H + (u,Tz)_H$, $\forall z \in \mathcal{D}(T)$.

As we know that $(Tz,u)=(z,T^* u)$, $\forall z \in \mathcal{D}(T)$ - because $u \in \mathcal{D}(T^*)$ - we obtain $\{x,y\} \perp G_T$, hence $\mathcal{V}\left(G_{T^*}\right) \subset (G_T)^{\perp}$. ■

We are now ready for:

Proof of Theorem 11.1 If the domain of T^* is not dense in H, its closure $\overline{\mathcal{D}(T^*)}$ is a strict subspace of H, hence, there exists an element $h \neq \mathbf{0}$ in H, such that $h \perp \mathcal{D}(T^*)$.

It follows that $\{\mathbf{0},h\} \perp \mathcal{V}\left(G_{T^*}\right)$; $\forall x \in \mathcal{D}(T^*)$; we have in fact $(\{0,h\}, \{-T^* x, x\})_{H\times H} = (h,x)_H = 0$, $\forall x \in \mathcal{D}(T^*)$.

Hence $\{\mathbf{0},h\} \in \left[\mathcal{V}\left(G_{T^*}\right)\right]^{\perp} = (G_T)^{\perp\perp}$ by (11.6). On the other hand (as previously seen) $(G_T)^{\perp\perp} = \overline{G_T}$ (closure in $H \times H$).

As T is closable, let $T \subset \tilde{T}$, with a closed operator $\tilde{T}$. Then $\overline{G_T} \subset \overline{G}_{\tilde{T}} = G_{\tilde{T}}$ and $\{\mathbf{0},h\} \in G_{\tilde{T}}$, so that $h = \tilde{T}\mathbf{0} = \mathbf{0}$, a contradiction. ■

Corollary *Let T be linear operator with dense domain in H which is also closable. Then the closure of T, $\overline{T}$ equals the second adjoint T^{**} (which exists by Theorem 11.1).*

Proof We have in fact:

$$G_{\overline{T}} = \overline{G_T} = (G_T)^{\perp\perp} = \left[\mathcal{V}\left(G_{T^*}\right)\right]^{\perp} = G_{T^{**}}$$

(by (11.5) applied to T^*). Thus $G_{\overline{T}} = G_{T^{**}}$ which means that $\overline{T} = T^{**}$.

In particular, if T *is closed with dense domain* H, *then* $T = T^{**}$. ∎

Note finally:

Proposition 11.1 *If T is a closable operator with dense domain in the Hilbert space H, the equality*

$$(\overline{T})^* = T^*$$

holds true.

In fact, from the above corollary we obtain $\overline{T} = T^{**}$, hence $(\overline{T})^* = T^{***}$. Now, $G_{T^{***}} = \left(\mathcal{V}G_{T^{**}}\right)^{\perp}$ by (11.5), and $\left(\mathcal{V}\left(G_{T^{**}}\right)\right)^{\perp} = \left(G_{T^*}\right)^{\perp\perp}$ (by (11.6)) $= G_{T^*}$ (T^* is always closed!). Thus, $T^{***} = T^* = (\overline{T})^*$. ∎

12. In this section we give a few more concrete examples related to various properties of the discontinuous linear operators previously discussed.

First let $\mathcal{H} = L^2(\mathbb{R})$, the Hilbert space of complex-valued square integrable functions on the real line.

Define $\mathcal{D} = \left\{\varphi \in L^2(\mathbb{R}) \text{ such that } x\varphi(x) \in L^2(\mathbb{R})\right\}$.

Thus $\mathcal{D}$ contains all continuous functions with compact support on $\mathbb{R}$, and is, accordingly, a dense subset of $\mathcal{H}$.

Then, define on $\mathcal{D}$ the operator T through the formula

$$(T\varphi)(x) = x\varphi(x), \quad \forall x \in \mathbb{R}, \ \forall \varphi \in \mathcal{D}. \tag{12.1}$$

Therefore T maps $\mathcal{D}$ into $L^2(\mathbb{R})$.

Consider the sequence $\varphi_n(\cdot)$ where

$$\varphi_n(x) = 1 \ \text{ for } \ n \le x \le n+1, \ \varphi_n(x) = 0 \ \text{ for } \ -\infty < x < n \ \text{ and } \ n+1 < x < \infty .$$

We see that

$$\|\varphi_n(\cdot)\|_{L^2(\mathbb{R})} = \left(\int_n^{n+1} \mathrm{d}x \right)^{\frac{1}{2}} = 1 , \qquad \forall n = 1,2,\ldots$$

and then

$$\|T\varphi_n\|_{L^2(\mathbb{R})} = \left(\int_n^{n+1} x^2 \,\mathrm{d}x \right)^{\frac{1}{2}} = \left[\frac{(n+1)^3}{3} - \frac{n^3}{3} \right]^{\frac{1}{2}} = \frac{1}{\sqrt{3}} \left[3n^2 + 3n + 1 \right]^{\frac{1}{2}}$$

so that

$$\lim_{n\to\infty} \|T\varphi_n\|_{\mathcal{H}} = +\infty .$$

The operator T is discontinuous.

Next, let $\mathcal{H} = L^2(\Omega)$, where Ω is a bounded 'regular' open set in $\mathbb{R}^n$, with boundary $S = \partial\Omega$.

Let $\mathcal{D}$ be composed of functions in $C^2(\overline{\Omega})$ which are zero on S. Define the operator $\Delta = \sum_{i=1}^{n} \frac{\partial^2}{\partial x_i^2}$ on $\mathcal{D}$, considered as a subspace of $L^2(\Omega)$.

It is a symmetric operator. In fact, by Green's formula we have

$$\int_\Omega (u\Delta v - v\Delta u) \,\mathrm{d}x = \int_S \left(u\frac{\partial v}{\partial n} - v\frac{\partial u}{\partial n} \right) \mathrm{d}\sigma = 0$$

for $u, v \in \mathcal{D}$, which obviously means that

$$(u, \Delta v)_{L^2(\Omega)} = (v, \Delta u)_{L^2(\Omega)} = (\Delta u, v)_{L^2(\Omega)}$$

(here we take $L^2(\Omega)$ over the real field $\mathbb{R}$).

Our last discussion here concerns the operator T in the first example. We show that $T = T^*$ (T is a self-adjoint discontinuous operator).

Let $\psi(\cdot) \in \mathcal{D}(T^*)$. Then, $\forall \varphi(\cdot) \in \mathcal{D}(T)$ we have $(T\varphi, \psi) = (\varphi, T^*\psi)$ which means that

$$\int_{\mathbb{R}} x\varphi(x)\overline{\psi}(x) \,\mathrm{d}x = \int_{\mathbb{R}} \varphi(x) \overline{(T^*\psi)(x)} \,\mathrm{d}x , \qquad \forall \varphi \in \mathcal{D}(T) .$$

Let us take the function $\varphi(\cdot)$ in $\mathcal{D}(T)$ of the form:

$$\varphi(x)=1\,, \quad \text{for} \quad \alpha \le x \le x_0\,, \qquad \varphi(x)=0\,, \quad \text{for} \quad x \notin [\alpha, x_0]\,.$$

It follows therefore that

$$\int_\alpha^{x_0} x\,\overline{\psi}(x)\,\mathrm{d}x = \int_\alpha^{x_0} \overline{(T^*\psi)(x)}\,\mathrm{d}x$$

which readily implies, a.e. in $x_0 \in \mathbb{R}$, the equality

$$x_0\,\overline{\psi}(x_0) = \overline{(T^*\psi)(x_0)}$$

which in turn implies that $x \to x\,\psi(x) \in L^2(\mathbb{R})$.

Therefore we obtained $\psi(\cdot) \in L^2(\mathbb{R})$ and $x\,\psi(x) \in L^2(\mathbb{R})$ which means that $\psi(\cdot) \in \mathcal{D}(T)$. Hence, $\mathcal{D}(T^*) \subset \mathcal{D}(T)$.

On the other hand, T is symmetric on $\mathcal{D}(T)$:

$$(T\varphi,\psi) = \int_{\mathbb{R}} x\varphi(x)\overline{\psi}(x)\,\mathrm{d}x = \int_{\mathbb{R}} \varphi(x)\,\overline{x\psi(x)}\,\mathrm{d}x = (\varphi, T\psi)$$

if φ, ψ both belong to $\mathcal{D}(T)$.

Hence, $\mathcal{D}(T) \subset \mathcal{D}(T^*)$. ■

13. We shall dedicate a few lines to a class of linear operators appearing in several applications: the so-called maximal accretive (also called 'maximal monotone') operators in Hilbert spaces.

Thus, let H be a Hilbert space, and A, $\mathcal{D}(A) \subset H \to H$ be a linear operator. We say that A is an accretive operator (or else, that $-A$ is a dissipative operator) if

$$\operatorname{Re}(Ah,h) \ge 0\,, \quad \forall h \in \mathcal{D}(A)\,. \tag{13.1}$$

We say also that A is a maximal accretive operator if (13.1) holds and also

$$R(I+A) = H \tag{13.2}$$

(R stands for 'range'; I is the identity operator in H; $\mathcal{D}(I+A) = \mathcal{D}(A)$).

We shall establish:

Proposition 13.1 *If* A, $\mathcal{D}(A) \subset H \to H$ *is a maximal accretive operator, then* $\mathcal{D}(A)$ *is dense in* H *and* A *is a closed operator.*

Proof In order to demonstrate the density of $\mathcal{D}(A)$ we shall take any $f \in H$ which is orthogonal to $\mathcal{D}(A)$ and show that it is the zero-vector.

Thus, let $f \in H$ and assume that $(f,h) = 0$, $\forall h \in \mathcal{D}(A)$. Now use (13.2), and find $h_0 \in \mathcal{D}(A)$ such that

$$(I+A)h_0 = f, \quad \text{that is} \quad h_0 + Ah_0 = f.$$

Then we obtain

$$0 = (h_0, f)_H = (h_0, h_0 + Ah_0) = \|h_0\|^2 + (h_0, Ah_0)$$

$$0 = (f, h_0) \ = (h_0 + Ah_0, h_0) = \|h_0\|^2 + (Ah_0, h_0)$$

We add both equalities, to obtain

$$0 = 2\|h_0\|^2 + 2\,\mathrm{Re}(Ah_0, h_0)\ ,$$

and because of (13.1), $h_0 = \mathbf{0}$. Therefore $f = (I+A)h_0 = \mathbf{0}$ too. Thus, $\overline{\mathcal{D}(A)} = H$.

Next, let us see why A is a closed operator.

First we show, completing (13.2), that, $\forall f \in H$, there exists a *unique* solution $u \in \mathcal{D}(A)$ to the equation

$$(I+A)u = f\ . \tag{13.3}$$

In fact, if $\bar{u} \in \mathcal{D}(A)$ is another solution to (13.3), we obtain $\bar{u} + A\bar{u} = f$. Then $(u-\bar{u}) + A(u-\bar{u}) = \mathbf{0}$ and

$$(u-\bar{u},\ u-\bar{u}) + (A(u-\bar{u}), u-\bar{u}) = 0\ ;$$

also

$$(u-\bar{u},\ u-\bar{u}) + (u-\bar{u},\ A(u-\bar{u})) = 0\ ;$$

summing we obtain

$$2\|u-\overline{u}\|^2+2\operatorname{Re}(A(u-\overline{u}),u-\overline{u})=0$$

and, again, because of (13.1), it follows that $u=\overline{u}$.

These arguments show the existence of the inverse operator

$$(I+A)^{-1}, \quad H\to \mathcal{D}(A).$$

We prove that it is a continuous operator and $\left\|(I+A)^{-1}\right\|_{\mathcal{L}(H)}\leq 1$. Consider the relation $u+Au=f$, where $f\in H,\ u\in \mathcal{D}(A)$. Taking scalar products one gets:

$$(u,u)+(Au,u)=(f,u)$$

$$(u,u)+(u,Au)=(u,f)$$

therefore

$$2\|u\|^2+2\operatorname{Re}(Au,u)=(f,u)+(u,f)\leq 2\|f\|\,\|u\| \tag{13.4}$$

and, consequently, because of (13.1)

$$\|u\|^2\leq\|f\|\,\|u\|, \quad \|u\|\leq\|f\|, \quad \left\|(I+A)^{-1}f\right\|\leq\|f\|, \quad \forall f\in H. \tag{13.5}$$

Next, the operator $(I+A)^{-1}$, $H\to\mathcal{D}(A)$ is continuous, hence closed (see Remark 2 to Proposition 9.1). Its inverse, which is the operator $I+A$, $\mathcal{D}(A)\to H$, will be closed too (see Proposition 10.1). Finally, if $I+A$ is closed then A is closed: in fact take a sequence (u_n) in $\mathcal{D}(A)$, where $u_n\to u_0$ while $Au_n\to v_0$.

It follows that

$$u_n\to u_0, \quad (I+A)u_n\to u_0+v_0,$$

whence

$$u_0\in\mathcal{D}(I+A)=\mathcal{D}(A), \quad \text{and} \quad (I+A)u_0=u_0+v_0,$$

that is $Au_0=v_0$. Thus, A *is closed.* ■

Also of interest is the following:

Proposition 13.2 *Let A, $\mathcal{D}(A)\subset H\to H$, be a symmetric maximal accretive operator. Then A is self-adjoint.*

Proof Let us consider the operator $\mathcal{J}=(I+A)^{-1}$ which is, as seen above, a linear continuous operator in H. We see now that $\mathcal{J}$ is symmetric, that is

$$(\mathcal{J}u,v)=(u,\mathcal{J}v)\ ,\qquad \forall u,v\in H\ . \tag{13.6}$$

In fact, if $u_1=\mathcal{J}u=(I+A)^{-1}u,\ v_1=\mathcal{J}v=(I+A)^{-1}v$, it follows that

$$u=(I+A)u_1\ ,\qquad v=(I+A)v_1\ .$$

Accordingly:

$$(\mathcal{J}u,v)=\left(u_1,v_1+Av_1\right)=\left(u_1,v_1\right)+\left(u_1,Av_1\right)$$

$$(u,\mathcal{J}v)=\left(u_1+Au_1,v_1\right)=\left(u_1,v_1\right)+\left(Au_1,v_1\right)\ .$$

As u_1,v_1 belongs to $\mathcal{D}(A)=\mathcal{D}(I+A)$ and A is symmetric on $\mathcal{D}(A)$, we derive

$$(\mathcal{J}u,v)=(u,\mathcal{J}v)\ ,\qquad \forall u,v\in H\ .$$

In order to prove $A=A^*$ it will be enough to see that $A^*\subset A$. Thus, let $u\in\mathcal{D}(A^*)$ and denote $f=u+A^*u$. We obtain:

$$(f,v)=(u+A^*u,v)=(u,v)+(A^*u,v)=(u,v)+(u,Av)\ , \tag{13.7}$$

$\forall v\in\mathcal{D}(A)$ (note that $(Av,u)=(v,A^*u)$ if $v\in\mathcal{D}(A)$ and $u\in\mathcal{D}(A^*)$).

Take now any $w\in H$; thus $\mathcal{J}w=(I+A)^{-1}w=v$ belongs to $\mathcal{D}(A)$. Introducing this in (13.7) one gets

$$(f,\mathcal{J}w)=(u,(I+A)v)=(u,w)$$

and, because of (13.6),

$$(\mathcal{J}f,w)=(u,w)\ ,\qquad \forall w\in H\ .$$

This implies that $u=\mathcal{J}f$, an element of $\mathcal{D}(A)$. ■

14. We shall briefly discuss in this section some rudiments of so-called 'spectral theory' for linear operators on Banach spaces (not necessarily continuous).

Thus, let T be a linear operator over the Banach space X which is taken with the field $\mathbb{K}$ (the real numbers or complex numbers).

The resolvent set of T, denoted $\rho(T)$, is composed of $\lambda \in \mathbb{K}$, such that $\lambda I - T$: $\mathcal{D}(T) \to X$ is bijective and $(\lambda I - T)^{-1} \in \mathcal{L}(X)$; thus:

$$(\lambda I - T)^{-1}(\lambda I - T)x = x, \quad \forall x \in \mathcal{D}(T)\ ,$$

$$(\lambda I - T)(\lambda I - T)^{-1}x = x, \quad \forall x \in X\ .$$

Note the following:

If $\rho(T) \neq \varnothing$ then T is a closed operator. (In fact, $\lambda I - T$ is closed, as the inverse of a continuous, everywhere defined linear operator; this implies T is closed: for, let (x_n) be a sequence in $\mathcal{D}(T)$, $x_n \to x_0$, $Tx_n \to y_0$. Then $(\lambda I - T)x_n \to \lambda x_0 - y_0$, and so, $x_0 \in \mathcal{D}(\lambda I - T) = \mathcal{D}(T)$, while $(\lambda I - T)x_0 = \lambda x_0 - y_0$, which gives $Tx_0 = y_0$; this argument has been used in the previous section.)

Note also that, if T is a closed operator, then

$$\rho(T) = \{\lambda \in \mathbb{K};\ \lambda I - T;\ \mathcal{D}(T) \to X \text{ is bijective}\} \tag{14.1}$$

(for, by the closed graph theorem, the closed operator $(\lambda I - T)^{-1}$ which is everywhere defined in X will be continuous).

In the definitions above I is the identity operator on X. We write sometimes $\lambda - T$ for $\lambda I \quad T$.

The operator $(\lambda - T)^{-1} = R_\lambda(T)$ is called the resolvent (operator) of T; this is also sometimes written $R(\lambda; T) = (\lambda I - T)^{-1}$.

The spectrum of T is, by definition, the set $\mathbb{K} - \rho(T) = \sigma(T)$.

The point spectrum of T is composed of eigenvalues of T; it is denoted by $\sigma_p(T)$. Thus

$$\sigma_p(T) = \{\lambda \in \mathbb{K}, \text{ such that } \exists x \in X,\ x \neq \mathbf{0}, \text{ and } Tx = \lambda x\}\ . \tag{14.2}$$

Note that if $\lambda \in \sigma_p(T)$ then $(\lambda I - T)$ is not injective, hence $\lambda \in \sigma(T)$.

We define also the concepts of the residual spectrum of T, $\sigma_r(T)$, through

$$\sigma_r(T) = \{\lambda \in \mathbb{K},\ \lambda I - T \text{ is injective and the range } R(\lambda I - T) \text{ is not dense in } X\}\ ; \tag{14.3}$$

the continuous spectrum of T, $\sigma_c(T)$, is defined by

$$\sigma_c(T) = \{\lambda \in \mathbb{K},\ \lambda I - T \text{ is injective, } R(\lambda I - T) \text{ is dense in } X,\ (\lambda I - T)^{-1} \text{ is not continuous}\}\ . \tag{14.4}$$

We may note the following:

Proposition 14.1 *For any linear operator* T, $\mathcal{D}(T) \subset X \to X$, *the four sets* $\rho(T)$, $\sigma_p(T)$, $\sigma_c(T)$, $\sigma_r(T)$ *are mutually exclusive, and we have*

$$\mathbb{K} = \rho(T) \cup \sigma_p(T) \cup \sigma_c(T) \cup \sigma_r(T) \tag{14.5}$$

$$\sigma_p(T) \cup \sigma_c(T) \cup \sigma_r(T) = \sigma(T)\ . \tag{14.6}$$

Proof If $\lambda \in \sigma_p(T)$, $\lambda - T$ is *not* injective, hence $\lambda \notin \sigma_r(T)$ and $\lambda \notin \sigma_c(T)$. Thus $\sigma_p(T) \cap \sigma_c(T) = \varnothing$, $\sigma_p(T) \cap \sigma_r(T) = \varnothing$. Then, if $\lambda \in \sigma_c(T)$, the range $R(\lambda I - T)$ is dense, which is not possible if $\lambda \in \sigma_r(T)$. Thus, $\sigma_c(T) \cap \sigma_r(T) = \varnothing$.

Next, if $\lambda \in \sigma_p(T)$, then $\lambda - T$ is not injective, hence $\lambda \in \sigma(T)$. If $\lambda \in \sigma_c(T)$, then $(\lambda I - T)^{-1}$ is not continuous, hence, again, $\lambda \in \sigma(T)$. If $\lambda \in \sigma_r(T)$, the range of $\lambda I - T$ is not dense, hence $\lambda I - T$ is not surjective, again $\lambda \in \sigma(T)$.

If $\lambda \in \sigma(T)$, then $\lambda \in C\rho(T)$. Remember that

$$\begin{aligned}\rho(T) &= \{\lambda \in \mathbb{K},\ \lambda - T \text{ is bijective and } (\lambda - T)^{-1} \text{ is continuous}\} \\ &= \{\lambda \in \mathbb{K},\ \lambda - T \text{ is injective, is surjective, and } (\lambda I - T)^{-1} \text{ is continuous}\}\ .\end{aligned}$$

Therefore, if $\lambda \notin \rho(T)$, then $\lambda - T$ is not injective, or $\lambda - T$ is not surjective, or $(\lambda - T)^{-1}$ is not continuous.

If $\lambda - T$ is not injective, then $\lambda \in \sigma_p(T)$, if $\lambda - T$ is not surjective, then $\lambda \in \sigma_r(T)$ or $\lambda \in \sigma_c(T)$ or even $\lambda \in \sigma_p(T)$.

If $(\lambda - T)^{-1}$ is not continuous, then $\lambda \in \sigma_r(T)$ or $\lambda \in \sigma_c(T)$. This discussion settles (14.6) (with disjoint components in the left-hand side).

As $\mathbb{K} = \rho(T) \cup \sigma(T)$, by definition of $\sigma(T)$, we get (14.5) too. ∎

The following result appears useful on several occasions.

Proposition 14.2 *(The resolvent identity) Let T be a closed linear operator on the Banach space X, and $\lambda, \mu \in \rho(T)$. Then*

$$R(\lambda,T)-R(\mu,T)=(\mu-\lambda)R(\lambda,T)R(\mu,T) \tag{14.7}$$

holds true.

Remark As we implicitly assume that $\rho(T)\neq\varnothing$, the restriction to closed operators is only apparent.

Also, (14.7) implies the commutativity property:

$$R(\lambda,T)R(\mu,T)=R(\mu,T)R(\lambda,T)\ , \qquad \forall\lambda,\mu\in\rho(T)\ . \tag{14.8}$$

Proof Note the equality, for $\lambda, \mu \in \rho(T)$:

$$\begin{aligned} R(\lambda,T) &= R(\lambda,T)(\mu-T)(\mu-T)^{-1} \\ &= R(\lambda,T)(\mu-\lambda+\lambda-T)(\mu-T)^{-1} \\ &= (\mu-\lambda)R(\lambda,T)R(\mu,T)+R(\mu,T) \end{aligned}$$

(remember that $R(\lambda,T)(\lambda-T)x=x$, $\forall x\in\mathcal{D}(T)$, while $(\lambda-T)R(\lambda,T)x=x$, $\forall x\in X$).

■

Let us now give a few concrete examples.

Example 1 Let $X=C[0,1]$ – the space of real-valued continuous functions on [0,1] with the sup-norm. Let T be the linear operator defined on continuously differentiable functions $[0,1]\to\mathbb{R}$, by the usual derivative:

$$(T\varphi)(t)=\varphi'(t)\ . \tag{14.9}$$

It has been indicated that T is a closed operator. Let us look for eigenvalues of T, which means solving the equation $\lambda\varphi-\varphi'=0$, $\forall\lambda\in\mathbb{R}$, with non-trivial solution.

We obtain: $\varphi(t)=ce^{\lambda t}$, $c\neq 0$.

Hence in the space X (over $\mathbb{R}$), $\sigma_p(T)=\mathbb{R}$. This also shows that $\sigma(T)=\mathbb{R}$, $\rho(T)=\varnothing$.

Remark If we take instead $X = C[0,1]$ composed of complex-valued continuous functions, which is a vector space over $\mathbb{C}$, and the same differentiation operator as above, we get $\sigma_p(T) = \mathbb{C}$, $\rho(T) = \varnothing$.

Example 2 Let, again, $X = C[0,1]$, the space of continuous complex-valued functions on $[0,1]$, and define $\mathcal{D} = \{$continuously differentiable functions on $[0,1]$ which are null for $t = 0\}$.

Then put T, $\mathcal{D} \to X$

$$T\varphi = \varphi'(t) \ .$$

Now the situation is completely different from the previous example and in fact, we shall see that $\rho(T) = \mathbb{C}$.

First we note that, $\forall \lambda \in \mathbb{C}$, the operator $\lambda - T$ is injective: if $\varphi(\cdot) \in \mathcal{D}$ and $\lambda\varphi - T\varphi = \mathbf{0}$, we get $\lambda\varphi(t) - \varphi'(t) = 0$ with $\varphi(0) = 0$; its solution (the only one) is $\varphi(t) \equiv 0$ (for, $\frac{\mathrm{d}}{\mathrm{d}t}\left(\mathrm{e}^{-\lambda t}\varphi(t)\right) = 0$ gives $\mathrm{e}^{-\lambda t}\varphi(t) = \varphi(0) = 0$).

Next, $\forall f \in C[0,1]$, the equation $(\lambda - T)\varphi = f$, $\varphi \in \mathcal{D}$, becomes

$$\lambda\varphi - \varphi' = f, \quad \varphi(0) = 0 \ .$$

Its unique solution is obtained by the formula

$$\varphi(t) = -\int_0^t \mathrm{e}^{\lambda(t-s)} f(s)\,\mathrm{d}s \ ,$$

as is readily seen. Therefore, the operator $\lambda I - T$, $\mathcal{D} \to C[0,1]$, is bijective, $\forall \lambda \in \mathbb{C}$. Let us estimate now the sup-norm of the solution φ in terms of the sup-norm of the given f.

We obtain

$$|\varphi(t)| = \mathrm{e}^{(\mathrm{Re}\,\lambda)t}\left|\int_0^t \mathrm{e}^{-\lambda s} f(s)\,\mathrm{d}s\right|$$

$$\le \mathrm{e}^{(\mathrm{Re}\,\lambda)t} \sup_{[0,1]} |f(s)| \int_0^t \mathrm{e}^{-(\mathrm{Re}\,\lambda)s}\,\mathrm{d}s$$

$$= e^{(\mathrm{Re}\,\lambda)t}\|f\|_X \frac{\left(1-e^{(-\mathrm{Re}\,\lambda)t}\right)}{\mathrm{Re}\,\lambda}$$

$$= \|f\|_X \frac{1}{\mathrm{Re}\,\lambda}\left(e^{(\mathrm{Re}\,\lambda)t}-1\right)$$

$$\leq C(\lambda)\|f\|_X$$

$\forall t \in [0,1]$, hence $\|\varphi(\cdot)\|_X = C(\lambda)\|f\|_X$, for $\mathrm{Re}\,\lambda \neq 0$.

If $\mathrm{Re}\,\lambda = 0$ we obtain

$$|\varphi(t)| \leq \|f\|_X \cdot t \leq \|f\|_X , \qquad \forall t \in [0,1],$$

hence $\|\varphi\|_X \leq \|f\|_X$.

These estimates show that $(\lambda I - T)^{-1}$, $C[0,1] \to \mathcal{D}$, is also in $\mathcal{L}(X)$, $\forall \lambda \in \mathbb{C}$. Thus, in fact, $\rho(T) = \mathbb{C}$, $\sigma(T) = \varnothing$.

Example 3 Let $X = C[0,1]$ over $\mathbb{R}$; $\mathcal{D} = \{$continuously differentiable functions, $[0,1] \to \mathbb{R}$, $\varphi(t)$, such that $\varphi(0) = \varphi(1)\}$ and let T, $\mathcal{D} \to \mathbb{R}$, be as previously, the differentiation operator:

$$T\varphi = \varphi' .$$

If we now look for eigenvalues, we have the differential equation

$$\lambda\varphi - \varphi' = 0 , \qquad \text{with} \qquad \varphi(0) = \varphi(1) .$$

As $\varphi(t) = Ce^{\lambda t}$, $\varphi(0) = C$, $\varphi(1) = Ce^{\lambda}$, $C = Ce^{\lambda}$, $\lambda = 0$. Hence $\varphi(t) = C$ is the eigenvector corresponding to $\lambda = 0$. We get $\sigma_p(T) = \{0\}$.

Considering instead $C[0,1]$ over $\mathbb{C}$ and the same operator, the equation $e^{\lambda} = 1$ gives $\lambda = \pm 2ni\pi$, $n = 0,1,2,\ldots$. Now $\sigma_p(T) = \{2ni\pi\}_{n\in\mathbb{Z}}$.

Example 4 If X is a finite dimensional vector space and $T \in \mathcal{L}(X)$, then, as is well-known, T is represented by a matrix $\left(t_{ij}\right)_{i,j=1}^{n}$ where $\left(t_{ij}\right)$ are complex numbers when X is over $\mathbb{C}$.

The eigenvalues of T now appear as the roots of the equation

$$\det\left(\lambda\delta_{ij} - t_{ij}\right) = 0 .$$

Example 5 Let $X = L^2(\mathbb{R})$, the space of real-valued, square-integrable functions on the real line.

Let $\mathcal{D} = \{\varphi(\cdot) \in L^2(\mathbb{R}), \text{ such that } x\varphi(x) \in L^2(\mathbb{R})\}$.

Define the operator T, $\mathcal{D} \to X$, by the formula

$$(T\varphi)(x) = x\varphi(x) \ . \tag{14.10}$$

Then, we shall see that $\sigma_c(T) = \mathbb{R}$.

In fact, first we show that, $\forall \lambda_0 \in \mathbb{R}$, the operator $\lambda_0 I - T$ is injective, $\mathcal{D} \to X$.

Take therefore $\varphi(\cdot) \in \mathcal{D}$, in such a way that $\lambda_0\varphi - T\varphi = \mathbf{0}$ in $L^2(\mathbb{R})$. This means that

$$\lambda_0\varphi(x) - x\varphi(x) = 0 \quad \text{a.e. on } \mathbb{R}, \qquad (\lambda_0 - x)\varphi(x) = 0 \quad \text{a.e.}$$

This is not possible unless $\varphi(x) = 0$ a.e. on $\mathbb{R}$.

Next, let us note that the domain of the inverse operator $(\lambda_0 - T)^{-1}$, that is the range of the operator $\lambda_0 - T$, contains all the functions $\psi(\cdot) \in L^2(\mathbb{R})$ which are zero in a neighborhood of λ_0.

For, let us take such a $\psi(\cdot)$ and consider the equation

$$(\lambda_0 - T)\varphi = \psi$$

which means $\lambda_0\varphi(x) - x\varphi(x) = \psi(x)$ a.e.

The solution (pointwise, a.e.)

$$\varphi(x) = \frac{1}{\lambda_0 - x}\psi(x) = \begin{cases} 0, & \text{for } |x - \lambda_0| < \delta \\ \dfrac{\psi(x)}{\lambda_0 - x}, & \text{for } |x - \lambda_0| > \delta \end{cases}$$

belongs to $L^2(\mathbb{R})$, in view of the obvious estimate

$$|\varphi(x)| \le \frac{1}{\delta}|\psi(x)| \ , \qquad \forall x \in \mathbb{R} \ .$$

Note also that the set of those $\psi(\cdot) \in L^2(\mathbb{R})$ which are zero in a (variable) neighborhood of λ_0 is dense in $L^2(\mathbb{R})$.

Thus, by now we know also that $R(\lambda_0 - T)$ is dense in X.

Our next step is to establish that the operator: $(\lambda_0 - T)^{-1}$ is not continuous.

In fact, its continuity would imply (see Proposition 2.2 in this chapter) that, for some $m > 0$, the lower bound

$$\|\lambda_0\varphi - T\varphi\|_{L^2} \geq m\|\varphi\|_{L^2}, \quad \forall \varphi \in \mathcal{D}(T) \tag{14.11}$$

holds true.

Consider now (in order to disprove (14.11)) a sequence of functions in $\mathcal{D}(T) = \mathcal{D}$, given by

$$\varphi_n(x) = \begin{cases} 1 & \text{for } |x - \lambda_0| < \alpha_n, \\ 0 & \text{for } |x - \lambda_0| \geq \alpha_n, \end{cases} \qquad \alpha_n > 0, \ \forall n \in \mathbb{N} .$$

Compute

$$\begin{aligned} &\|(\lambda_0 - T)\varphi_n\|_{L^2(\mathbb{R})} \\ &= \left(\int_{\mathbb{R}} |\lambda_0 \varphi_n(x) - x\varphi_n(x)|^2 \, \mathrm{d}x \right)^{\frac{1}{2}} \\ &= \left(\int_{\lambda_0 - \alpha_n}^{\lambda_0 + \alpha_n} (\lambda_0 - x)^2 \, \mathrm{d}x \right)^{\frac{1}{2}} \\ &= \left(\frac{2\alpha_n^3}{3} \right)^{\frac{1}{2}} . \end{aligned}$$

Also, $\|\varphi_n\|_{L^2(\mathbb{R})} = \left(\int_{\lambda_0 - \alpha_n}^{\lambda_0 + \alpha_n} \mathrm{d}x \right)^{\frac{1}{2}} = \sqrt{2\alpha_n}$.

Inequality (14.11) would mean, if $\varphi = \varphi_n$,

$$\left(\frac{2}{3} \right)^{\frac{1}{2}} \alpha_n^{3/2} \geq m\sqrt{2}\alpha_n^{\frac{1}{2}}, \quad \forall n \in \mathbb{N},$$

for some $m > 0$, or also

$$\frac{1}{\sqrt{3}}\alpha_n \geq m > 0, \quad \forall n \in \mathbb{N}.$$

This is false if $\alpha_n \to 0$.

Remark It follows that $\sigma(T)=\sigma_c(T)=\mathbb{R},\ \rho(T)=\varnothing$.

Our last example:

Example 6 Let $X=l^2$ – the space of real-valued sequences (ξ_k) such that $\sum_1^\infty \xi_k^2<\infty$.

Define the operator T by

$$T(\xi_1,\xi_2,\dots\xi_n,\dots)=(0,\xi_1,\xi_2,\dots)\ ; \tag{14.12}$$

it is obviously a linear continuous operator on l^2.

We shall see that $\lambda=0$ belongs to $\sigma_r(T)$; first, the operator $-T$ is injective, as is readily seen.

Next, the range of the operator $-T$ is formed of 'square-integrable' sequences $(\eta_k)_1^\infty$, such that $\eta_1=0$.

This last set is *not* dense in l^2; for, take $(\xi_1,\xi_2\dots)$ in l^2, and then a sequence $\left(\eta_k^p\right)_{k=1}^\infty$ where $\eta_1^p=0,\ \forall p\in\mathbb{N}$, and $\sum_{k=1}^\infty\left(\eta_k^p-\xi_k\right)^2\to 0$ as $p\to\infty$.

It follows that $\left(\eta_1^p-\xi_1\right)^2\to 0$ as $p\to\infty$, hence $\xi_1=0$.

We now terminate this section with the following:

Theorem 14.1 *Let H be a Hilbert space over $\mathbb{C}$ and A, $\mathcal{D}(A)\subset H\to H$ be a self-adjoint operator. Then, if $\lambda\in\mathbb{C}$ and $\operatorname{Im}\lambda\neq 0$, it follows that $\lambda\in\rho(A)$ and that*

$$\|R(\lambda,A)\|=\left\|(\lambda-A)^{-1}\right\|_{\mathcal{L}(H)}\le\frac{1}{|\operatorname{Im}\lambda|}\ . \tag{14.13}$$

Proof First note that A is a symmetric operator, hence $(Ah,h)=(h,Ah),\ \forall h\in\mathcal{D}(A)$, which shows that the function (Ah,h) is real-valued, and consequently, that all eigenvalues of A are real.

Now, take $\lambda\in\mathbb{C}$ and $h\in\mathcal{D}(A)$; we obtain

$$\left|((\lambda-A)h,h)_H\right|\le\|(\lambda-A)h\|\,\|h\|$$

by the Cauchy-Schwarz inequality.

Furthermore, we have

$$((\lambda - A)h, h) = \lambda\|h\|^2 - (Ah, h) ,$$

hence

$$\operatorname{Im}((\lambda - A)h, h) = (\operatorname{Im}\lambda)\|h\|^2 .$$

On the other hand

$$|\operatorname{Im}((\lambda - A)h, h)| \le \|(\lambda - A)h\| \, \|h\| ,$$

whence

$$|(\operatorname{Im}\lambda)| \, \|h\|^2 \le \|(\lambda - A)h\| \, \|h\| ,$$

and

$$\|(\lambda - A)h\| \ge |\operatorname{Im}\lambda| \, \|h\| , \qquad \forall h \in \mathcal{D}(A) . \tag{14.14}$$

We find that the inverse operator $(\lambda - A)^{-1}$ exists and is continuous when $\operatorname{Im}\lambda \neq 0$.

Next, we shall prove that $(\lambda - A)^{-1}$ is everywhere defined in H (when $\operatorname{Im}\lambda \neq 0$). As a first step we show that $(\lambda - A)^{-1}$ is densely defined in H. Otherwise, $\exists k \in H$, $k \neq \mathbf{0}$, and $k \perp (\lambda - A)h$, $\forall h \in \mathcal{D}(A)$. This means that $(\lambda h - Ah, k)_H = 0$, $\forall h \in \mathcal{D}(A)$, which can also be written as

$$(Ah, k) = (\lambda h, k) , \qquad \forall h \in \mathcal{D}(A) .$$

This last equality entails: $k \in \mathcal{D}(A^*)$, and $A^* k = \overline{\lambda} k$. However, our operator A is self-adjoint, so that $A^* = A$ and we have therefore $Ak = \overline{\lambda} k$, with $k \neq \mathbf{0}$.

Furthermore, $\operatorname{Im}\overline{\lambda} = -\operatorname{Im}\lambda \neq 0$, which means that $\overline{\lambda}$ is a non-real eigenvalue of the symmetric operator A, which of course is not possible. Next step: the operator $(\lambda - A)^{-1}$ is closed and densely defined.

Take any $h \in H$ and then a sequence $x_n \in \mathcal{D}\big((\lambda - A)^{-1}\big)$, such that $x_n \to h$. As $(\lambda - A)^{-1}$ is continuous (hence uniformly continuous) on its domain, it follows that $(\lambda - A)^{-1} x_n$ has a limit k in H.

From the definition of closed operators we then infer that: $h \in \mathcal{D}\big((\lambda - A)^{-1}\big)$ and $(\lambda - A)^{-1} h = k$.

The continuous operator $(\lambda - A)^{-1}$ is therefore everywhere defined in H; we obtain $\lambda \in \rho(A)$. The inequality (14.13) is a consequence of (14.14). ∎

15. In this section we introduce the concept of dual operators in Banach space, especially for linear, not necessarily continuous, operators.

Thus, let E, F be Banach spaces; A, $\mathcal{D}(A) \subset E \to F$ be a linear operator, defined on $\mathcal{D}(A)$ in E with range in F. We shall assume here that

$$\mathcal{D}(A) \text{ is dense in } E\,. \tag{15.1}$$

Consider next spaces E' and F' – the complete normed spaces composed of linear continuous functionals on E (resp. F), with norm as given in (3.6) of Chapter I.

Define now a set $\mathcal{D}'$ in F' as follows:

$$\mathcal{D}' = \left\{ f \in F', \text{ such that } \exists C > 0 \text{ with: } |f(Au)| \le C\|u\|_E, \forall u \in \mathcal{D}(A) \right\}\,. \tag{15.2}$$

We see without difficulty that $\mathcal{D}'$ is a linear set (subspace) in F'.

Let K be the (common) field of E and F. Define the function g, $\mathcal{D}(A) \to K$, by

$$g(u) = f(Au) \quad (\text{where } f \in \mathcal{D}')\,. \tag{15.3}$$

Note that g is a linear function ($g = f \circ A$, the composition of two linear functions). Furthermore, as we assumed $f \in \mathcal{D}'$, it follows that

$$|g(u)| \le C\|u\|_E\,, \quad \forall u \in \mathcal{D}(A)\,.$$

Thus, g is Lipschitz (hence uniformly) continuous on $\mathcal{D}(A)$, and using (15.1) we derive that g has a unique linear continuous extension, denoted by $\tilde{g}$, $E \to K$. The inequality

$$|\tilde{g}(u)| \le C\|u\|_E\,, \quad \forall u \in E \tag{15.4}$$

holds true.

Therefore $\tilde{g} \in E'$. Now, we define, for all $f \in \mathcal{D}'$

$$A'f = \tilde{g}\,; \quad \text{hence } \mathcal{D}(A') = \mathcal{D}'\,. \tag{15.5}$$

We can summarize what is being discussed by

$$\text{if} \quad f \in \mathcal{D}(A') \text{ and } u \in \mathcal{D}(A), \quad \text{then} \qquad f(Au) = g(u) = \tilde{g}(u) = (A'f)(u)\,. \tag{15.6}$$

Note that A maps $\mathcal{D}(A) \subset E$ into F, while A' maps $\mathcal{D}(A') = \mathcal{D}' \subset F'$ into E'.

Let us demonstrate:

Proposition 15.1 *The operator A', $\mathcal{D}' \to E'$, is a linear closed operator.*

Proof

(a) Let $f_1, f_2 \in \mathcal{D}'$. We have - using (15.6) - the equalities:

$$f_1(Au) = (A'f_1)(u)\,, \quad f_2(Au) = (A'f_2)(u)\,, \qquad \forall u \in \mathcal{D}(A)\,. \tag{15.7}$$

Therefore we obtain, by addition

$$(f_1 + f_2)(Au) = (A'f_1 + A'f_2)(u)\,, \qquad \forall u \in \mathcal{D}(A)\,. \tag{15.8}$$

On the other hand, $f_1 + f_2 \in \mathcal{D}'$ (which is a linear subspace of F'), hence, from (15.6), we can also write

$$(f_1 + f_2)(Au) = (A'(f_1 + f_2))(u)\,, \qquad \forall u \in \mathcal{D}(A)\,.$$

Consequently, we see that

$$A'(f_1 + f_2)(u) = (A'f_1 + A'f_2)(u)\,, \qquad \forall u \in \mathcal{D}(A)\,. \tag{15.9}$$

Now, $\mathcal{D}(A)$ is dense in E and $A'(f_1 + f_2)$, $A'f_1$, $A'f_2$ all belong to E'. Therefore (15.9) extends to all $u \in E$ which means that

$$A'(f_1 + f_2) = A'f_1 + A'f_2\,, \qquad \forall f_1, f_2 \in \mathcal{D}' = \mathcal{D}(A')\,. \tag{15.10}$$

In a similar way we get $A'(\lambda f) = \lambda A'(f)$, $\forall \lambda \in K$, $\forall f \in \mathcal{D}'$.

(b) Let us show why A' is a closed operator. Take therefore a sequence (f_n) in $\mathcal{D}'$, such that $f_n \to f_0$ (in F') while $A'f_n \to g_0$ in E'. We also have the equalities

$$f_n(Au) = (A'f_n)(u)\,, \qquad \forall u \in \mathcal{D}(A),\ n = 1,2,\dots . \tag{15.11}$$

As convergence in F', E' implies 'pointwise' convergence, we derive from (15.11), as $n \to \infty$, the equality

$$f_0(Au) = g_0(u)\,, \qquad \forall u \in \mathcal{D}(A)\,.$$

Therefore,

$$|f_0(Au)| = |g_0(u)| \le \|g_0\|_{E'} \|u\|_E\,, \qquad \forall u \in \mathcal{D}(A)$$

and (15.2) then gives $f_0 \in \mathcal{D}(A')$ and accordingly

$$f_0(Au) = (A'f_0)(u) = g_0(u)\,, \qquad \forall u \in \mathcal{D}(A)\,.$$

Again, as $\mathcal{D}(A)$ is dense in E, we get $A'f_0 = g_0$. ■

Also of interest is the following:

Proposition 15.2 *Let us define*

$$\begin{aligned} \mathcal{D}_1' = \{ & f \in F' \text{ such that } \exists F \in E' \text{ with equality} \\ & f(Au) = F(u), \quad \forall u \in \mathcal{D}(A)\}\,. \end{aligned} \tag{15.12}$$

$$A_1' f = F\,, \qquad \forall f \in \mathcal{D}_1'\,. \tag{15.13}$$

Then $\mathcal{D}_1' = \mathcal{D}'$ *and* $A_1' = A'$.

Proof Note first that for $f \in \mathcal{D}_1'$, the element $F \in E'$ is uniquely determined; again this is because $\mathcal{D}(A)$ is dense in E.

Next, if $f \in \mathcal{D}'$ we have seen previously the existence of $\tilde{g} \in E'$ such that $f(Au) = \tilde{g}(u)$, $\forall u \in \mathcal{D}(A)$. Thus, $f \in \mathcal{D}_1'$.

Conversely, if $f \in \mathcal{D}_1'$, we have

$$|f(Au)| = |F(u)| \le \|F\|_{E'} \|u\|_E$$

so that $f \in \mathcal{D}'$ too.

We see also that $F = \tilde{g}$ hence $A_1' = A'$. ■

Example 1 Consider the special case where A is linear continuous $E \to F$ (thus, $\mathcal{D}(A) = E$).

We note, for any $f \in F'$, the inequality

$$|f(Au)| \le \|f\|_{F'} \|Au\|_F \le \|f\|_{F'} \|A\|_{\mathcal{L}(E,F)} \|u\|_E \ , \qquad \forall u \in E \ .$$

Therefore, the set $\mathcal{D}'$ in (15.2) is now equal to F', and

$$f(Au) = (f \circ A)(u) \ , \qquad \forall u \in E \ .$$

Hence $f \circ A \in E'$ and we get

$$(f \circ A)(u) = (A'f)(u) \ , \qquad \forall u \in E, \ \forall f \in F' ,$$

that is

$$A'f = f \circ A \ . \tag{15.14}$$

This last equality shows that $A' \in \mathcal{L}(F', E')$ and also that

$$\|A'f\|_{E'} \le \|f\|_{F'} \|A\|_{\mathcal{L}(E,F)} \ , \qquad \forall f \in F' \tag{15.15}$$

hence

$$\|A'\|_{\mathcal{L}(F',E')} \le \|A\|_{\mathcal{L}(E,F)} \ . \tag{15.16}$$

Example 2 Let E, F both be Hilbert spaces and $A \in \mathcal{L}(E,F)$. The adjoint operator $A^* \in \mathcal{L}(F,E)$ has been given by the equality

$$(Ah, k)_F = (h, A^* k)_E \ , \qquad \forall h \in E, \ \forall k \in F \tag{15.17}$$

(see (3.1) in this chapter).

On the other hand, by the above discussion, the dual operator $A' \in \mathcal{L}(F', E')$, is defined by

$$(A'f)(u) = f(Au) \ , \qquad \forall u \in E, \ \forall f \in F' \ . \tag{15.18}$$

We shall now make use of Theorem 3.4 in Chapter 1 and the conjugate linear mapping σ from the dual H' of a Hilbert space H onto H, $\forall f \in H'$, $\exists! y_f \in H$, such that $f(x) = (x, y_f)$, $\forall x \in H$, and then $\sigma(f) = y_f$. Thus we have

$$f(x) = (x, \sigma(f))_H \,, \qquad \forall x \in H \,. \tag{15.19}$$

In our present situation we obtain

$$f(Au) = (Au, \sigma_{F'}(f))_F \,, \qquad \forall f \in F', \ \forall u \in E \tag{15.20}$$

and also

$$(A'f)(u) = (u, \sigma_{E'}(A'f))_E \,, \qquad \forall f \in F', \ \forall u \in E \,. \tag{15.21}$$

Therefore, from (15.18) we derive

$$(Au, \sigma_{F'}(f))_F = (u, \sigma_{E'}(A'f))_E \,, \qquad \forall u \in E, \ \forall f \in F' \,. \tag{15.22}$$

Let us denote $\sigma_{F'}(f) = \tilde{f} \in F$; then $f = \sigma_{F'}^{-1}\tilde{f}$ and

$$(Au, \tilde{f})_F = \left(u, \sigma_{E'} A' \sigma_{F'}^{-1} \tilde{f}\right)_E \,, \qquad \forall u \in E, \ \forall \tilde{f} \in F \,. \tag{15.23}$$

If we compare with (15.17), we find the equality

$$A^* \tilde{f} = \sigma_{E'} A' (\sigma_{F'})^{-1} \tilde{f} \,, \qquad \forall \tilde{f} \in F \tag{15.24}$$

(in fact in (15.24), $\sigma_{E'}$ acts from E' onto E while $(\sigma_{F'})^{-1}$ from F onto F').

Equality (15.24) establishes an important relationship between adjoint and dual operators in pairs of Hilbert spaces.

Chapter III

Duality: the Hahn-Banach extension theorem

Introduction

In this chapter we discuss some further results pertaining to duality theory in Banach spaces: extension of linear continuous functionals defined on linear subspaces, dual or second dual spaces, and related concepts.

1. The extension theorem of linear functionals has already been considered for the special case of Hilbert spaces as Theorem 5.1 in Chapter I.

Here we put ourselves in more general situations.

Theorem 1.1 *Let X be a vector space over $\mathbb{R}$, and p, $X \to \mathbb{R}$ be a convex function: $p(\alpha x+(1-\alpha)y) \leq \alpha p(x)+(1-\alpha)p(y)$, $\forall x, y \in X$, $\forall \alpha \in [0,1]$.*

Next, consider a linear subspace Y of X and a linear function λ, $Y \to \mathbb{R}$, which satisfies the inequality

$$\lambda(x) \leq p(x)\,, \qquad \forall x \in Y\,. \tag{1.1}$$

Then, there exists a linear function Λ, $X \to \mathbb{R}$, with the properties

$$\Lambda(x) \leq p(x)\,, \quad \forall x \in X\,, \qquad \Lambda(x) = \lambda(x)\,, \quad \forall x \in Y\,. \tag{1.2}$$

The *proof* will only be sketched.

Let $z \in X$, $z \notin Y$. Define the linear space $\widetilde{Y}$ by

$$\widetilde{Y} = Y + \{\alpha z\}_{\alpha \in \mathbb{R}} = \{y+\alpha z\}_{y \in Y}\,, \qquad \alpha \in \mathbb{R}\,. \tag{1.3}$$

Next, one establishes the following inequality:

$$\sup_{y \in Y, \alpha > 0} \frac{1}{\alpha}[-p(y-\alpha z)+\lambda(y)] \leq \inf_{y \in Y, \alpha > 0} \frac{1}{\alpha}[p(y+\alpha z)-\lambda(y)]\,. \tag{1.4}$$

We may therefore conclude the existence of a real number a, such that

$$\sup_{y\in Y,\alpha>0} \frac{1}{\alpha}[-p(y-\alpha z)+\lambda(y)] \le a \le \inf_{y\in Y,\alpha>0} \frac{1}{\alpha}[p(y+\alpha z)-\lambda(y)] \tag{1.5}$$

and then we define a function $\tilde{\lambda}$, $\tilde{Y} \to \mathbb{R}$, by

$$\tilde{\lambda}(y+\alpha z) = \lambda(y)+\alpha\cdot a\,, \qquad \forall y \in Y,\ \forall \alpha \in \mathbb{R}\ . \tag{1.6}$$

This function $\tilde{\lambda}$ is linear as is readily seen.

It also satisfies the inequality $\tilde{\lambda}(x) \le p(x)$, $\forall x \in \tilde{Y}$, which is in fact

$$\lambda(y)+\alpha\cdot a \le p(y+\alpha z)\,, \qquad \forall y \in Y,\ \forall \alpha \in \mathbb{R}\ . \tag{1.7}$$

Take first $\alpha > 0$. In this case (1.7) is equivalent to

$$a \le \frac{1}{\alpha}[p(y+\alpha z)-\lambda(y)]\,, \qquad \forall y \in Y$$

which follows from (1.5).

Next, if $\alpha < 0$, (1.7) is equivalent to

$$a \ge \frac{1}{\alpha}[p(y+\alpha z)-\lambda(y)]\ ;$$

if we put $\alpha' = -\alpha$, we get

$$a \ge -\ \frac{1}{\alpha'}[p(y-\alpha' z)-\lambda(y)] = \frac{1}{\alpha'}[-p(y-\alpha' z)+\lambda(y)]$$

which follows from the left-hand side of (1.5).

The above construction indicates how to obtain a linear extension of λ (with conservation of (1.1)) 'one dimension' at a time.

The remaining part of the proof is based on Zorn's lemma (Theorem 4.1, Chapter I).

The maximal element guaranteed by that theorem will be to a linear extension Λ, $X \to \mathbb{R}$, such that (1.2) holds true.

Precisely, let us denote by $\mathcal{E}$ the family of all the linear extensions of λ from Y to a greater linear subspace of X, and verifying (1.1) on their domain of definition. For e_1, e_2 in $\mathcal{E}$ we say that $e_1 \prec e_2$ if $\mathcal{D}(e_1)$ (domain of e_1) is contained in $\mathcal{D}(e_2)$ and if $e_2(x) = e_1(x)$, $\forall x \in D(e_1)$. The relation '$\prec$' is an ordering on $\mathcal{E}$.

Next, if $\{e_\alpha\}$ is a totally ordered subset in $\mathcal{E}$, with $\mathcal{D}(e_\alpha) = X_\alpha$ we construct an upper bound e for $\{e_\alpha\}$, taking $\mathcal{D}(e) = \bigcup X_\alpha$, and $e(x) = e_\alpha(x)$ if $x \in X_\alpha$.

Therefore, $\exists \Lambda$, an element which is maximal in $\mathcal{E}$; thus, $\mathcal{D}(\Lambda) = X^1 \subset X$ and $\Lambda(x) = \lambda(x)$, $\forall x \in Y$, $\Lambda(x) \le p(x)$, $\forall x \in X^1$.

Now, if $X^1 \subset X$ strictly we can produce (as in the first part of this proof) an extension of Λ on a subspace X^2 containing X^1, therefore contradicting the maximality of Λ. ■

Our next step is to establish a similar result for vector spaces over $\mathbb{C}$. We state:

Theorem 1.2 *Let X be a vector space over $\mathbb{C}$ and p, $X \to \mathbb{R}^+$ be a function such that: $p(\alpha x + \beta y) \le |\alpha| p(x) + |\beta| p(y)$, $\forall x, y \in X$, $\forall \alpha, \beta \in \mathbb{C}$, with $|\alpha| + |\beta| = 1$.*

Next, consider a linear subspace Y of X and a linear function λ, $Y \to \mathbb{C}$, which satisfies the inequality

$$|\lambda(x)| \le p(x)\,, \qquad \forall x \in Y\,. \tag{1.8}$$

Then, there exists a linear function Λ, $X \to \mathbb{C}$, with the properties

$$|\Lambda(x)| \le p(x)\,, \quad \forall x \in X\,, \qquad \Lambda(x) = \lambda(x)\,, \quad \forall x \in Y\,.$$

Proof We first consider the function l, $Y \to \mathbb{R}$, which is defined by

$$l(x) = \operatorname{Re} \lambda(x)\,, \qquad \forall x \in Y\,. \tag{1.9}$$

We note that $l(x)$ is a linear function, $Y \to \mathbb{R}$ (for instance, if $\mu \in \mathbb{R}$, then $l(\mu x) = \operatorname{Re} \lambda(\mu x) = \operatorname{Re}[\mu \lambda(x)] = \mu \operatorname{Re} \lambda(x)$).

Note also that for $\mathrm{i} = \sqrt{-1}$

$$l(\mathrm{i}x) = \operatorname{Re}[\lambda(\mathrm{i}x)] = \operatorname{Re}[\mathrm{i}\lambda(x)] = -\operatorname{Im} \lambda(x)\,.$$

We obtain the representation formula

$$\lambda(x) = l(x) - \mathrm{i}l(\mathrm{i}x) \ . \tag{1.10}$$

Our assumption on p entails, obviously, that p is a convex function on Y. We are thus entitled to apply Theorem 1.1 in order to obtain a linear function L, $X \to \mathbb{R}$, such that

$$L(x) = l(x) \ , \quad \forall x \in Y \qquad \text{and} \qquad L(x) \le p(x) \ , \quad \forall x \in X \ . \tag{1.11}$$

Next, we define the function Λ, from X to $\mathbb{C}$, by the formula

$$\Lambda(x) = L(x) - \mathrm{i}L(\mathrm{i}x) \ , \quad \forall x \in X \ . \tag{1.12}$$

It will be the function that we are looking for. In fact: if $x \in Y$, then

$$\Lambda(x) = l(x) - \mathrm{i}l(\mathrm{i}x) = \lambda(x) \quad \text{by (1.10)} \ ;$$

$$\Lambda(x+y) = L(x+y) - \mathrm{i}L(\mathrm{i}x+\mathrm{i}y) = L(x) + L(y) - \mathrm{i}L(\mathrm{i}x) - \mathrm{i}L(\mathrm{i}y) = \Lambda(x) + \Lambda(y) \ ;$$

for $\alpha \in \mathbb{R}$,

$$\Lambda(\alpha x) = L(\alpha x) - \mathrm{i}L(\mathrm{i}\alpha x) = \alpha L(x) - \mathrm{i}\alpha L(\mathrm{i}x) = \alpha\Lambda(x) \ ;$$

$$\Lambda(\mathrm{i}x) = L(\mathrm{i}x) - \mathrm{i}L(-x) = L(\mathrm{i}x) + \mathrm{i}L(x) = \mathrm{i}\Lambda(x) \ .$$

These relations show that $\Lambda = \lambda$ on Y and that Λ is linear (in X over $\mathbb{C}$). Next, note that if $\alpha \in \mathbb{C}$, $|\alpha| = 1$, then $p(\alpha x) \le p(x)$; also $p(x) = p\left(\frac{1}{\alpha}\alpha x\right) \le \frac{1}{|\alpha|} p(\alpha x) = p(\alpha x)$. Therefore $p(\alpha x) = p(x)$ if $|\alpha| = 1$.

On the other hand, if $\Lambda(x) = 0$, then $|\Lambda(x)| \le p(x)$; if $\Lambda(x) \ne 0$, put $\theta = \operatorname{Arg}\Lambda(x)$ and note that $\operatorname{Re}\Lambda(x) = L(x)$ (from (1.12)). Then $\Lambda(x) = |\Lambda(x)|\mathrm{e}^{\mathrm{i}\theta}$, hence $|\Lambda(x)| = \Lambda(x)\mathrm{e}^{-\mathrm{i}\theta}$ $= \Lambda\left(\mathrm{e}^{-\mathrm{i}\theta}x\right) = \operatorname{Re}\Lambda\left(\mathrm{e}^{-\mathrm{i}\theta}x\right) = L\left(\mathrm{e}^{-\mathrm{i}\theta}x\right) \le p\left(\mathrm{e}^{-\mathrm{i}\theta}x\right) = p(x)$. ■

Of special importance to us is the following extension theorem in normed vector spaces.

Theorem 1.3 *Let X be a normed vector space over $\mathbb{R}$ or $\mathbb{C}$, Y, a linear subspace of X, and λ, a linear continuous function, $Y \to \mathbb{R}$ or $Y \to \mathbb{C}$. Then, there exists $\Lambda \in X'$, such that $\Lambda = \lambda$ on Y and $\|\Lambda\|_{X'} = \|\lambda\|_{Y'}$.*

Proof

(a) Assume X is linear over $\mathbb{R}$; put, $\forall x \in X$, $p(x) = \|\lambda\|_{Y'} \cdot \|x\|$. Then, $\forall \alpha \in [0,1]$, we have, for $x, y \in X$

$$\begin{aligned} p(\alpha x + (1-\alpha)y) &= \|\lambda\|_{Y'} \cdot \|\alpha x + (1-\alpha)y\| \\ &\leq \|\lambda\|_{Y'}(\alpha\|x\| + (1-\alpha)\|y\|) \\ &= \alpha p(x) + (1-\alpha)p(y) . \end{aligned} \tag{1.13}$$

As $\lambda \in Y'$ we have $|\lambda(x)| \leq \|\lambda\|_{Y'} \cdot \|x\|$, $\forall x \in Y$ which means that $|\lambda(x)| \leq p(x)$, $\forall x \in Y$.

According to Theorem 1.1, there exists Λ, a linear function $X \to \mathbb{R}$, such that

$$\Lambda(x) \leq \|\lambda\|_{Y'}\|x\| , \quad \forall x \in X .$$

Then

$$-\Lambda(x) = \Lambda(-x) \leq \|\lambda\|_{Y'}\|x\| , \quad \forall x \in X ;$$

therefore

$$|\Lambda(x)| \leq \|\lambda\|_{Y'}\|x\| , \quad \forall x \in X .$$

This entails

$$\Lambda \in X' , \quad \|\Lambda\|_{X'} \leq \|\lambda\|_{Y'} . \tag{1.14}$$

On the other hand, using $\Lambda = \lambda$ on Y, one gets:

$$\|\Lambda\|_{X'} = \sup_{\substack{x \in X \\ \|x\|=1}} |\Lambda(x)| \geq \sup_{\substack{x \in Y \\ \|x\|=1}} |\Lambda(x)| = \sup_{\substack{x \in Y \\ \|x\|=1}} |\lambda(x)| = \|\lambda\|_{Y'} . \tag{1.15}$$

Combining (1.14)-(1.15) we get the equality of the norms.

(b) Assume X is linear over $\mathbb{C}$; then, again, take $p(x) = \|\lambda\|_{Y'} \cdot \|x\|$. If $\alpha, \beta \in \mathbb{C}$ we get

$$\begin{aligned} p(\alpha x+\beta y) &= \|\lambda\|_{Y'}\|\alpha x+\beta y\| \\ &\le \|\lambda\|_{Y'}(|\alpha|\,\|x\|+|\beta|\,\|y\|) \\ &= |\alpha|p(x)+|\beta|p(y)\ . \end{aligned}$$

Thus, we can apply Theorem 1.2 and terminate the proof as in (a) above. ■

We next point out the following:

Corollary 1.1 *Let X be a vector normed space, and $x_0 \in X$, $x_0 \neq \mathbf{0}$. Then, there exists $F \in X'$, such that*

$$F(x_0)=\|x_0\|\ , \quad \|F\|_{X'}=1\ . \tag{1.16}$$

In fact, let $Y=\{\alpha x_0\}$ where $\alpha \in K$ (the field of *X*). Define $f;\ Y \to K$, by the formula

$$f(\alpha x_0)=\alpha\|x_0\|\ . \tag{1.17}$$

Then *f* is linear on *Y*, $|f(\alpha x_0)|=|\alpha|\,\|x_0\|=\|\alpha x_0\|$, $f(x_0)=\|x_0\|$. Therefore, $\|f\|_{Y'}=1$.

Thus, the extension of *f* to the whole *X*, whose existence follows from Theorem 1.3, is the *F* in (1.16).

Note This corollary shows that if $x_0, y_0 \in X$ and $F(x_0)=F(y_0)$, $\forall F \in X'$, then $x_0=y_0$.

Example Let (x_n) be a sequence in *X*, and $x_0 \in X$ too.

Assume that

$$F(x_n) \to F(x_0)\ , \quad \forall F \in X' \tag{1.18}$$

(we say that $x_n \to x_0$ weakly in *X*).

Then the weak limit x_0 is uniquely determined.

2. We saw in previous sections that, if *X* is any vector normed space over the field *K* ($\mathbb{R}$ or $\mathbb{C}$), the dual space X' is defined as $X'=\mathcal{L}(X,K)$, and is a complete linear normed space, where

$$\|F\|_{X'} = \sup_{\substack{x\in X \\ \|x\|\le 1}} |F(x)| \,, \qquad \forall F \in X' \,.$$

It is often useful and/or interesting to consider the space $(X')' = X''$ which is also called the bidual space to X. At this point it also becomes convenient to use the notation x, x', x'' for elements of X, resp X', resp X''.

Thus, for instance, we have

$$\|x''\|_{X''} = \sup_{\|x'\|_{X'}\le 1} |x''(x')| \,. \tag{2.1}$$

An important observation is the following:

Any element x of X generates an element ϑx of X'' through the formula

$$(\vartheta x)(x') = x'(x) \,, \qquad \forall x' \in X' \,. \tag{2.2}$$

In fact, note first that, for $\alpha, \beta \in K$, $x', y' \in X'$, we have

$$\begin{aligned}(\vartheta x)(\alpha x' + \beta y') &= (\alpha x' + \beta y')(x) \\ &= \alpha x'(x) + \beta y'(x) \\ &= \alpha(\vartheta x)(x') + \beta(\vartheta x)(y') \,.\end{aligned} \tag{2.3}$$

(ϑx is a linear function, $X' \to K$.)

Next, we see the estimate

$$|(\vartheta x)(x')| = |x'(x)| \le \|x'\|_{X'} \|x\|_X \,, \qquad \forall x' \in X' \tag{2.4}$$

whence ϑx is also a continuous function, $X' \to K$.

(This last estimate also shows that

$$\|\vartheta x\|_{X''} \le \|x\|_X \qquad (\forall x \in X) \,.) \tag{2.5}$$

Looking again at (2.2), we realize that we have a mapping ϑ from X into X'', which is *injective*: if $\vartheta x = \vartheta y$ we derive that $(\vartheta x)(x') = (\vartheta y)(x')$, $\forall x' \in X'$, which means that

$$x'(x) = x'(y) \,, \qquad \forall x' \in X' \tag{2.6}$$

and, as seen in Section 1, (2.6) entails $x = y$.

Another important property of the mapping $\mathcal{J}$ is that it is a linear mapping:

We have in fact

$$\begin{aligned}(\mathcal{J}(x_1 + x_2))(x') &= x'(x_1 + x_2) \\ &= x'(x_1) + x'(x_2) \\ &= (\mathcal{J}x_1)(x') + (\mathcal{J}x_2)(x') \\ &= (\mathcal{J}x_1 + \mathcal{J}x_2)(x')\ , \quad \forall x_1, x_2 \in X,\ \forall x' \in X';\end{aligned}$$

thus $\mathcal{J}(x_1 + x_2) = \mathcal{J}x_1 + \mathcal{J}x_2$ and $\mathcal{J}$ is an additive mapping.

Then, $\forall \alpha \in K,\ \forall x \in X,\ \forall x' \in X'$,

$$(\mathcal{J}(\alpha x))(x') = x'(\alpha x) = \alpha x'(x) = \alpha(\mathcal{J}x)(x')$$

hence $\mathcal{J}(\alpha x) = \alpha \mathcal{J}x,\ \forall x \in X,\ \forall \alpha \in K$.

A consequence is that the image $\mathcal{J}(X)$ of X under $\mathcal{J}$ is a linear subspace of the bidual space X''.

If $\mathcal{J}(X) = X''$ we say *that X is a reflexive space.*[5] Note also that

$$\|\mathcal{J}x\|_{X''} = \|x\|_X\ , \quad \forall x \in X \tag{2.7}$$

($\mathcal{J}$ is an isometric mapping).

In fact, we apply Corollary 1.1, and find, for any $x \in X,\ x \neq \mathbf{0}$, an element x' of X', such that $\|x'\| = 1,\ x'(x) = \|x\|$.

Then we get:

$$(\mathcal{J}x)(x') = \|x\| = |(\mathcal{J}x)(x')|\ . \tag{2.8}$$

Using (2.5) we get: $\sup\limits_{\|x'\| \le 1} |(\mathcal{J}x)(x')| = \|\mathcal{J}x\|_{X''} = \|x\|_X$, that is (2.7).

Example 1 If X is a reflexive Banach space, then X' is also a reflexive B-space.

The assumption can be translated as follows: $\forall x'' \in X''$, there exists $x \in X$, such that $\mathcal{J}x = x''$, that is - from (2.2)

[5] In this case, as X'' is always a Banach space and $\mathcal{J}$ is isometric, it follows that X is complete, hence also a Banach space.

$$x''(x') = x'(x) , \quad \forall x' \in X' . \tag{2.9}$$

We must then prove that $\forall x''' \in X'''$ there exists $x' \in X'$ such that

$$x'''(x'') = x''(x') , \quad \forall x'' \in X'' . \tag{2.10}$$

Now, $\forall x'' \in X''$ we find $x \in X$ such that (2.9) holds; therefore $x'' = \vartheta x$, and then (2.10) becomes

$$x'''(\vartheta x) = (\vartheta x)(x') , \quad \forall x \in X . \tag{2.11}$$

Using also (2.2), we see that (2.11) is

$$x'''(\vartheta x) = x'(x) , \quad \forall x \in X, \tag{2.12}$$

with x' as unknown element.

This equation obviously has the solution $x' = x''' \cdot \vartheta$. ■

Example 2 (*Every Hilbert space is reflexive*) Let H be a Hilbert space and H', H'' its dual and bidual spaces. We shall use Theorem 3.4, Chapter I and therefore we have two conjugate linear isometric mappings σ_1, $H' \to H$ and σ_2, $H'' \to H'$: $\forall h' \in H'$, $\exists! \sigma_1(h') \in H$, such that

$$h'(h) = \left(h, \sigma_1(h')\right)_H , \quad \forall h \in H \tag{2.13}$$

and $\forall h'' \in H''$, $\exists! \sigma_2(h'') \in H'$, such that

$$h''(h') = \left(h', \sigma_2(h'')\right)_{H'} , \quad \forall h' \in H' . \tag{2.14}$$

Let us now use also equality (3.27) in Chapter I to get

$$\left(h', \sigma_2(h'')\right)_{H'} = \left(\sigma_1\left(\sigma_2(h'')\right),\ \sigma_1(h')\right)_H = h'\left(\sigma_1\left(\sigma_2(h'')\right)\right) \tag{2.15}$$

by (2.13).

Thus, from (2.14)-(2.15) we obtain $h''(h') = h'\left(\left(\sigma_1 \cdot \sigma_2\right)(h'')\right)$, $\forall h' \in H'$. Now $\left(\sigma_1 \cdot \sigma_2\right)(h'') = h \in H$ and we can say that $\forall h'' \in H''\ \exists h \in H$, such that

$$h''(h') = h'(h), \quad \forall h' \in H'.$$

H is therefore reflexive.

Lastly, we prove:

Example 3 Any *weakly bounded* set in a Banach space is also (strongly) bounded.

First, the (obvious) definition: A set A in X (a Banach space) is said to be weakly bounded if the numerical set

$$\{x'(x),\ x \in A\}$$

is bounded $\forall x' \in X'$.

Consider the ('canonical') mapping $\mathcal{J}$ previously introduced, so that

$$(\mathcal{J}x)(x') = x'(x), \quad \forall x \in X,\ \forall x' \in X'.$$

From weak boundedness we get, $\forall x' \in X'$, a constant $C_{x'} > 0$, such that

$$|x'(x)| \leq C_{x'}, \quad \forall x \in A. \tag{2.16}$$

This reads then

$$|(\mathcal{J}x)(x')| \leq C_{x'}, \quad \forall x \in A. \tag{2.17}$$

Consider the family $\{\mathcal{J}x\}_{x \in A}$ composed of elements in X''.

It is bounded on any $x' \in X'$, hence, by the uniform boundedness theorem, $\exists C > 0$, such that

$$\|\mathcal{J}x\|_{X''} \leq C, \quad \forall x \in A. \tag{2.18}$$

On the other hand, we also have $\|\mathcal{J}x\|_{X''} = \|x\|_X$, therefore

$$\|x\| \leq C, \quad \forall x \in A. \tag{2.19}$$

■

Chapter IV

Some singular (implicit) abstract differential equations

Introduction

In this chapter we introduce some classes of differential equations for functions with range in a Banach space: the equations are linear, of the form

$$Bu'(t) = Au(t) \tag{0.1}$$

where $u(t)$ is a function from a real interval I into a Banach space X, $u'(t)$ is its 'strong' derivative, and A and B are linear operators in X.

We define the so-called uniformly well-posed Cauchy problem for (0.1), and associate the Cauchy problem with a semidynamical system = semigroup of linear operators on X.

1. Thus, the main object in the study of differential equations in Banach space (which we often designate by the more concise name of 'abstract differential equations'), is a function $x(t)$, defined on a finite or infinite interval of the real line, I, with range in X. The 'usual' definitions of (strong) continuity or differentiability apply:

$x(t)$ is continuous in $t_0 \in I$ if $\forall \varepsilon > 0$, $\exists \delta(\varepsilon) > 0$ such that $t \in I$ and $|t - t_0| < \delta \Rightarrow \|x(t) - x(t_0)\| < \varepsilon$.

It is continuous in I if it is continuous $\forall t_0 \in I$. $x(t)$ admits a right-derivative in $t_0 \in I$ if it is defined in some interval $t_0 \le t < t_1$ and if $\exists y_0 \in X$ such that

$$\left\| \frac{1}{h}\left[x(t_0 + h) - x(t_0)\right] - y_0 \right\| \to 0 \quad \text{as} \quad h > 0 \text{ and } h \to 0 .$$

One writes in this case, $x'_+(t_0) = y_0$.

In a similar manner one defines the left-differentiability and derivative at t_0; if it exists it will be denoted by $x'_-(t_0)$.

If both the right and left derivatives exist at $t_0 \in I$, and if they are equal, we say that $x(t)$ is differentiable ('has a derivative' for $t = t_0$): $x'(t_0) = x'_+(t_0) = x'_-(t_0)$.

Therefore we get, from this discussion:

$$\lim_{h \to 0} \left\| \frac{1}{h} \left[x(t_0 + h) - x(t_0) \right] - x'(t_0) \right\| = 0 \tag{1.1}$$

where $t_0 \in I$ is interior to I.

If the function $x(t)$ has a derivative $\forall t_0 \in I$ which is interior to I, then $x(t)$ is continuous in t_0.

In fact, $\forall \varepsilon' > 0$, $\exists \delta(\varepsilon') > 0$ such that

$$\left\| \frac{1}{h} \left[x(t_0 + h) - x(t_0) \right] - x'(t_0) \right\| < \varepsilon' \quad \text{for} \quad 0 < |h| < \delta(\varepsilon') ;$$

then

$$\| x(t_0 + h) - x(t_0) - hx'(t_0) \| < \varepsilon' |h| \quad \text{for} \quad 0 < |h| < \delta(\varepsilon') .$$

Now, we also have

$$x(t_0 + h) - x(t_0) = x(t_0 + h) - x(t_0) - hx'(t_0) + hx'(t_0) ,$$

hence

$$\| x(t_0 + h) - x(t_0) \| \le \varepsilon' |h| + |h| \, \| x'(t_0) \| = |h| \left(\varepsilon' + \| x'(t_0) \| \right) < \varepsilon$$

if

$$0 < |h| < \frac{\varepsilon}{\varepsilon' + \| x'(t_0) \|} \quad \text{and} \quad 0 < |h| < \delta(\varepsilon') ,$$

$\forall \varepsilon > 0$ given in advance.

In the following we use the notation $C(I;X)$ for the (linear) space composed of all continuous functions $x(\cdot)$, $I \to X$.

Let us consider now two linear closed operators A and B, defined on $\mathcal{D}(A)$, $\mathcal{D}(B)$, subsets of X, with range in X. We assume also

$$\mathcal{D}(A) \cap \mathcal{D}(B) \quad \text{is dense in} \quad X . \tag{1.2}$$

Consider functions $u(\cdot)$, $[0,\infty) \to \mathcal{D}(A) \cap \mathcal{D}(B)$ with the following properties: $(Bu)(\cdot) \in C([0,\infty);X)$, $u'(t)$ exists $\forall t \in [0,\infty)$ (right-derivative for $t = 0!$), and $u'(t) \in \mathcal{D}(B)$ $\forall t \geq 0$; the equality

$$Bu'(t) = Au(t)\ , \qquad 0 \leq t < \infty \tag{1.3}$$

holds true.

(Note that it follows that $u(\cdot) \in C([0,\infty);X)$.)

Definition 1.1 The Cauchy problem for equation (1.3) on $[0,\infty)$: given $u_0 \in \mathcal{D}(A) \cap \mathcal{D}(B)$, find a function $u(\cdot)$ with the previously indicated properties, such that

$$u(0) = u_0\ . \tag{1.4}$$

An important special class of Cauchy problems consists of uniformly well-posed problems:

The Cauchy problem (1.3)-(1.4) is said to be uniformly well posed if:

$\forall u_0 \in \mathcal{D}(A) \cap \mathcal{D}(B)$ there exists one and only one solution $u(\cdot)$ on $[0,\infty)$; furthermore, if $u_{0,n} \in \mathcal{D}(A) \cap \mathcal{D}(B)$, $\forall n \in \mathbb{N}$ and $\lim_{n\to\infty} u_{0,n} = \mathbf{0}$, then, for the corresponding solutions $u_n(\cdot)$ such that $u_n(0) = u_{0,n}$, we have

$$\lim_{n\to\infty} u_n(t) = \mathbf{0} \tag{1.5}$$

uniformly in any finite interval $[0,T]$, $T < +\infty$.

Next, let us define the family of operators $U(t)$, $\mathcal{D}(A) \cap \mathcal{D}(B)$ into itself, depending on the parameter $t \geq 0$, precisely:

$$U(t)u_0 = u(t)\ , \tag{1.6}$$

the solution of (1.3) with condition (1.4).

We see, without much difficulty, the following properties:

(i) $U(0) = I$ (1.7)

the identity mapping of $\mathcal{D}(A) \cap \mathcal{D}(B)$; as $U(0)u_0 = u(0) = u_0 \ \forall u_0 \in \mathcal{D}(A) \cap \mathcal{D}(B)$.

(ii) $\lim_{\substack{t \to 0 \\ t > 0}} U(t)x = x \,, \quad \forall x \in \mathcal{D}(A) \cap \mathcal{D}(B)$ (1.8)

as $U(t)x = u(t)$, the solution $u(\cdot)$ with $u(0) = x$; the continuity of $u(\cdot)$ for $t = 0$ gives the result.

(iii) The mapping $U(t)$, $\mathcal{D}(A) \cap \mathcal{D}(B)$ into itself, is a linear mapping, $\forall t \geq 0$.
Take in fact $u_1, u_2 \in \mathcal{D}(A) \cap \mathcal{D}(B)$; then $u_1 + u_2 \in \mathcal{D}(A) \cap \mathcal{D}(B)$ and $U(t)(u_1 + u_2)$ is the solution of (1.3) with $u_1 + u_2$ as initial value; consider also the functions $u_1(t) = U(t)u_1$, $u_2(t) = U(t)u_2$, solutions of (1.3) with u_1, respectively u_2, as initial values; their sum $u_1(t) + u_2(t)$ is again a solution of the linear equation (1.3), with initial value $u_1 + u_2$. Using uniqueness of the Cauchy problem, it follows that

$$U(t)(u_1 + u_2) = U(t)u_1 + U(t)u_2 \,, \quad \forall t \geq 0, \ u_1, u_2 \in \mathcal{D}(A) \cap \mathcal{D}(B) \,.$$

In the same way one proves that $U(t)(\lambda x) = \lambda U(t)x$, $\forall \lambda \in K$ (the field of X), $\forall x \in \mathcal{D}(A) \cap \mathcal{D}(B)$, $\forall t \geq 0$.

(iv) The mapping $U(t)$ is continuous, $\mathcal{D}(A) \cap \mathcal{D}(B) \to \mathcal{D}(A) \cap \mathcal{D}(B)$, $\forall t \geq 0$.
Take in fact any sequence (x_n) in $\mathcal{D}(A) \cap \mathcal{D}(B)$ such that $x_n \to x_0 \in \mathcal{D}(A) \cap \mathcal{D}(B)$. Then $x_n - x_0 \to \mathbf{0}$ and the solution $U(t)(x_n - x_0) = U(t)x_n - U(t)x_0 \to \mathbf{0}$ ($\forall t \geq 0$), as $n \to \infty$. Thus, $U(t)x_n \to U(t)x_0$, $\forall t \geq 0$.

(v) (Semigroup property of $U(t)$) For any $t_1, t_2 \in [0, \infty)$, the equality

$$U(t_1 + t_2)x = U(t_1)U(t_2)x = U(t_2)U(t_1)x \,, \quad \forall x \in \mathcal{D}(A) \cap \mathcal{D}(B) \tag{1.9}$$

holds true.

In fact, let $u(t) = U(t)x$; then, for a fixed $a > 0$, consider the translated function $w_a(t)$, $t \geq 0 \to \mathcal{D}(A) \cap \mathcal{D}(B)$, which is defined by

$$w_a(t) = u(t + a) \,.$$

We see that $Bw'_a(t) = Bu'(t+a) = Au(t+a) = Aw_a(t)$, $\forall t \geq 0$.

Also, $w_a(0) = u(a) = U(a)x$ and $w_a(t) = U(t+a)x$. Consider also the function $z(t) = U(t)(U(a)x)$, which is the solution of (1.3) satisfying the condition $z(0) = U(a)x$. Therefore $z(0) = w_a(0)$, and both $z(\cdot)$ and $w_a(\cdot)$ are solutions of (1.3). The uniqueness of the Cauchy problem then entails $z(t) = w_a(t)$ $\forall t \geq 0$, which means $U(t)U(a)x = U(t+a)x$, $\forall t, a \in [0, \infty)$, $\forall x \in \mathcal{D}(A) \cap \mathcal{D}(B)$.

(vi) Extension of $U(t)$ from $\mathcal{D}(A) \cap \mathcal{D}(B)$ to the whole space X.

We need the following (interesting in itself).

Theorem 1.1 *Let E be a Banach space and T be a linear continuous operator, $\mathcal{D}(T) \subset E \to E$, where $\mathcal{D}(T)$ - the domain of T - is a linear dense subspace of E. Then, there exists one and only one linear continuous operator $\tilde{T}$, $E \to E$, such that $\tilde{T}x = Tx$, $\forall x \in \mathcal{D}(T)$.*

Remark We have already encountered this kind of result (in a special case) when we proved Theorem 5.1, Chapter I.

Proof of Theorem 1.1 First remember (see Theorem 3.1, Chapter I) that T is continuous, $D(T) \to E$, which implies that T is Lipschitz continuous on $\mathcal{D}(T)$

$$\|Tx - Ty\|_E \leq C\|x - y\|, \quad \forall x, y \in \mathcal{D}(T). \tag{1.10}$$

Next, given any $x \in E$, there exists a sequence (x_j) in $\mathcal{D}(T)$, where $x_j \to x$. Then (Tx_j) is a Cauchy sequence in E, and accordingly, $\exists y \in E$, $Tx_j \to y$. We now define $\tilde{T}x = y$, and we must prove that the element y does not depend on the sequence (x_j) which is convergent to x. Take therefore another sequence (x'_j) in $\mathcal{D}(T)$, such that $x'_j \to x$, and find $y' \in E$, such that $Tx'_j \to y'$. Now, the alternate sequence $x_1, x'_1, x_2, x'_2, \ldots, x_j, x'_j, \ldots$ is convergent to x, hence $\exists z \in E$, such that

$$Tx_1, Tx'_1, Tx_2, Tx'_2, \ldots, Tx_j, Tx'_j, \ldots \to z .$$

As (Tx_j), (Tx'_j) are both subsequences of this last sequence, we find that $y = y' = z$.

Thus, we now have $\tilde{T}$, $E \to E$, such that $\tilde{T}x = Tx$, $\forall x \in \mathcal{D}(T)$. This extended mapping $\tilde{T}$ is linear on E (for instance, if $x_1, x_2 \in E$ and if $(x_{1,n})$ is a sequence in $\mathcal{D}(T)$ with $x_{1,n} \to x_1$, while $(x_{2,n})$ is a sequence in $\mathcal{D}(T)$ with $x_{2,n} \to x_2$, then $x_{1,n} + x_{2,n} \to x_1 + x_2$ and we get

$$\tilde{T}(x_1 + x_2) = \lim T(x_{1,n} + x_{2,n}) = \lim(Tx_{1,n} + Tx_{2,n}) = \tilde{T}x_1 + \tilde{T}x_2 \text{).} \tag{1.11}$$

It remains to see the continuity of $\tilde{T}$: we know the estimate

$$\|Tx\| \le \|T\| \, \|x\| , \qquad \forall x \in \mathcal{D}(T) . \tag{1.12}$$

Hence, if $x \in E$ and (x_j) in $\mathcal{D}(T)$ is convergent to x, we obtain

$$\tilde{T}x = \lim Tx_j \; ; \qquad \|\tilde{T}x\| = \lim \|Tx_j\| \le \|T\| \lim \|x_j\| = \|T\| \, \|x\|$$

which gives:

$$\|\tilde{T}\| \le \|T\| , \qquad \tilde{T} \text{ is continuous, } E \to E. \tag{1.13}$$

On the other hand, for $x \in \mathcal{D}(T)$ we have

$$\|Tx\| = \|\tilde{T}x\| \le \|\tilde{T}\| \, \|x\| , \qquad \text{so that } \|T\| \le \|\tilde{T}\| .$$

The result of all this is that $\tilde{T}$ is continuous, $E \to E$ and $\|\tilde{T}\|_{\mathcal{L}(E,E)} = \|T\|_{\mathcal{L}(\mathcal{D}(T);E)}$. ■

We now return to (VI). Fix any $t \ge 0$ and apply Theorem 1.1 to the mapping $U(t)$ which is continuous, with dense domain $\mathcal{D}(A) \cap \mathcal{D}(B)$. We find a mapping $\tilde{U}(t)$, $X \to X$, again continuous ($\forall t \ge 0$) (it is the only continuous mapping $X \to X$ which coincides with $U(t)$ on $\mathcal{D}(A) \cap \mathcal{D}(B)$).

Furthermore - a very important fact! - $\tilde{U}(t)$ has again the semigroup property:

$$\tilde{U}(t_1 + t_2) = \tilde{U}(t_1)\tilde{U}(t_2) = \tilde{U}(t_2)\tilde{U}(t_1) , \qquad \forall t_1, t_2 \in [0, \infty) . \tag{1.14}$$

This is established as follows. Take any $x \in X$ and then a sequence (x_j) in $\mathcal{D}(A) \cap \mathcal{D}(B)$, such that $x_j \to x$. We have

$$\tilde{U}(t_1+t_2)x = \lim U(t_1+t_2)x_j = \lim U(t_1)\left[U(t_2)x_j\right]. \tag{1.15}$$

Now, as $x_j \to x$, $U(t_2)x_j \to \tilde{U}(t_2)x$ and then $U(t_1)\left[U(t_2)x_j\right] \to \tilde{U}(t_1)\left[\tilde{U}(t_2)x\right]$. Therefore we get

$$\tilde{U}(t_1+t_2)x = \tilde{U}(t_1)\tilde{U}(t_2)x, \quad \forall x \in X. \tag{1.16}$$

Thus, we have now associated to our uniformly well-posed Cauchy problem for the implicit equation (1.3) a family of operators $\tilde{U}(t)$, $0 \le t < +\infty$, acting from X into itself such that

$$\tilde{U}(t)u_0 = u(t), \text{ solution of (1.3) with } u(0) = u_0, \ \forall u_0 \in \mathcal{D}(A) \cap \mathcal{D}(B). \tag{1.17}$$

$$\tilde{U}(0)x = x, \ \forall x \in \mathcal{D}(A) \cap \mathcal{D}(B), \text{ hence } \forall x \in X; \ \tilde{U}(0) = I_X. \tag{1.18}$$

$$\tilde{U}(t)x \to x \text{ as } t \to 0, \ \forall x \in \mathcal{D}(A) \cap \mathcal{D}(B). \tag{1.19}$$

$\tilde{U}(t)$ is a linear continuous mapping, $X \to X$, $\forall t \ge 0$.

$\tilde{U}(t)$ is a, so-called, one-parameter semigroup of linear continuous operators, $X \to X$ (it satisfies (1.16) above).

We now prove some other properties of the family $\tilde{U}(t)$. To start with, we have:

Proposition 1.1 *For any $T > 0$ there exists $M_T > 0$, such that*

$$\left\|\tilde{U}(t)\right\|_{\mathcal{L}(X)} \le M_T, \quad \text{for } 0 \le t \le T. \tag{1.20}$$

Proof We first establish the following: $\forall T > 0$, $\exists M_T > 0$ such that

$$\|U(t)x\|_X \le M_T \|x\|_X, \quad \forall x \in \mathcal{D}(A) \cap \mathcal{D}(B), \ t \in [0,T]; \tag{1.21}$$

in fact, otherwise, $\exists T_0 > 0$ such that $\forall M > 0$, $\exists t_M \in \left[0, T_0\right]$, and $x_M \in \mathcal{D}(A) \cap \mathcal{D}(B)$, in such a way that

$$\|U(t_M)x_M\|_X > M\|x_M\|_X \tag{1.22}$$

holds true (hence $x_M \neq \mathbf{0}$).

Take $M = 1,2,3,\ldots$; we obtain sequences (t_n) in $[0,T_0]$ and (x_n) in $\mathcal{D}(A)\cap\mathcal{D}(B)$ with $x_n \neq \mathbf{0}$, $\forall n \in \mathbb{N}$, and

$$\|U(t_n)x_n\|_X > n\|x_n\|_X , \quad \forall n \in \mathbb{N} . \tag{1.23}$$

Define: $y_n = \dfrac{1}{n\|x_n\|} x_n$; thus, $y_n \to \mathbf{0}$ (as $n \to \infty$) and (from (1.23)),

$$\|U(t_n)y_n\| > 1 , \quad \forall n \in \mathbb{N} . \tag{1.24}$$

Consider also $u_n(t) = U(t)y_n$, a solution to $Bu_n'(t) = Au_n(t)$ on $[0,\infty)$, such that $u_n(0) = y_n$. As $y_n \to \mathbf{0}$ it should follow (from the uniform well-posedness of the Cauchy problem), that

$$\sup_{0\le t\le T_0} \|u_n(t)\| \to 0 \quad (n \to \infty) . \tag{1.25}$$

This however is impossible, since

$$\sup_{[0,T_0]} \|u_n(t)\| \ge \|u_n(t_n)\| > 1 . \tag{1.26}$$

Therefore we now have (1.21).

To end the proof of Proposition 1.1, take $x \in X$ and then a sequence (x_p) in $\mathcal{D}(A)\cap\mathcal{D}(B)$ such that $x_p \to x$. We have, by (1.21),

$$\|U(t)x_p\|_X \le M_T\|x_p\|_X , \quad \forall p \in \mathbb{N},\ 0 \le t \le T . \tag{1.27}$$

As $p \to \infty$ we readily get:

$$\|\tilde{U}(t)x\|_X \le M_T\|x\|_X , \quad 0 \le t \le T$$

which is (1.20).

The following result is important

Proposition 1.2 *We have*

$$\tilde{U}(t)x \in C([0,\infty);X)\,, \qquad \forall x \in X\,. \tag{1.28}$$

Take in fact (x_n) in $\mathcal{D}(A) \cap \mathcal{D}(B)$, $x_n \to x$. Then

$$\left\|\tilde{U}(t)x - \tilde{U}(t)x_n\right\|_X \le M_T \left\|x - x_n\right\|_X\,, \qquad \forall t \in [0,T],\ n \in \mathbb{N}\,.$$

Therefore, $\lim\limits_{n\to\infty} \tilde{U}(t)x_n = \tilde{U}(t)x$, uniformly on $[0,T]$.

All functions $\tilde{U}(t)x_n = U(t)x_n$ are solutions of the Cauchy problem, hence they belong to $C([0,\infty);X)$. Accordingly, $\tilde{U}(t)x \in C([0,\infty);X)$ too.

Remark This last property implies:

$$\lim_{t\to 0} \tilde{U}(t)x = x\,, \qquad \forall x \in X\,, \tag{1.29}$$

which complements property (1.19) above.

We are at this point able to establish a 'regularity' result for solutions of the Cauchy problem.

Proposition 1.3 *For the uniformly well-posed Cauchy problem (1.3)-(1.4), any solution* $u(t)$ *has a continuous derivative on* $[0,\infty)$ *(we say that* $u(\cdot) \in C^1([0,\infty);X)$*).*

(Remember that - in our definition of the solution - the (strong) derivative $u'(t)$ *exists* $\forall t \ge 0$ *but its continuity is not assumed.)*

Proof Using (1.6) we see the 'representation formula' for any solution $u(t)$:

$$u(t) = U(t)u_0 = U(t)u(0) \tag{1.30}$$

(because of (1.4)).

Next, if one uses (1.17) we see also that

$$u(t)=\widetilde{U}(t)u(0)\,, \qquad \forall t\geq 0\,. \tag{1.31}$$

If we compute the derivative of $u(t)$ using (1.31) we get, for $h>0$:

$$\frac{1}{h}[u(t+h)-u(t)]$$

$$=\frac{1}{h}[\widetilde{U}(t+h)u(0)-\widetilde{U}(t)u(0)] \;=\; \text{(by (1.14))}$$

$$\frac{1}{h}[\widetilde{U}(t)\widetilde{U}(h)u(0)-\widetilde{U}(t)u(0)]$$

$$=\widetilde{U}(t)\frac{1}{h}[\widetilde{U}(h)u(0)-u(0)]$$

$$=\widetilde{U}(t)\frac{1}{h}[u(h)-u(0)]\,.$$

Taking the limit as $h\to 0$, we obtain

$$\lim_{h\to 0}\frac{1}{h}[u(t+h)-u(t)]=\widetilde{U}(t)u'_{+}(0) \tag{1.32}$$

(remember that $\widetilde{U}(t)\in\mathcal{L}(X)$, $\forall t\geq 0$).

Therefore, we now have a representation formula for the derivative of $u(\cdot)$:

$$u'(t)=\widetilde{U}(t)u'(0) \tag{1.33}$$

which, in combination with (1.28), ends the proof. ■

Further regularity results are proved under certain supplementary assumptions:

Proposition 1.4 *For the uniformly well-posed Cauchy problem (1.3)-(1.4) and assuming the supplementary condition*

$$\begin{gathered}\widetilde{U}(t)x\in\mathcal{D}(B)\,, \qquad \forall x\in\mathcal{D}(B),\ \forall t\geq 0\\ \textit{and}\\ B\widetilde{U}(t)x=\widetilde{U}(t)Bx\,, \qquad \forall x\in\mathcal{D}(B),\ \forall t\geq 0\end{gathered} \tag{1.34}$$

it follows that

$$\tilde{U}(t)Ax = A\tilde{U}(t)x\,, \qquad \forall x \in \mathcal{D}(A) \cap \mathcal{D}(B) \tag{1.35}$$

and the functions $(Au)(\cdot)$, $(Bu')(\cdot)$ *both belong to* $C([0,\infty);X)$.

Proof First note that, for $x \in \mathcal{D}(A) \cap \mathcal{D}(B)$, we have $\tilde{U}(t)x = U(t)x = u(t)$, the solution with $u(0) = x$. Therefore $\tilde{U}(t)x \in \mathcal{D}(A) \cap \mathcal{D}(B)$ and $A\tilde{U}(t)x$ makes sense. Next, we see the equalities:

$$A\tilde{U}(t)x = AU(t)x = Au(t) = Bu'(t) = B(\tilde{U}(t)u'(0)) \tag{1.36}$$

(by (1.33)). Actually, $u'(0) \in \mathcal{D}(B)$ and from (1.34) it follows that

$$A\tilde{U}(t)x = \tilde{U}(t)(Bu'(0)) = \tilde{U}(t)Au(0) = \tilde{U}(t)Ax\,. \tag{1.37}$$

Again, as $\tilde{U}(t)z$ is continuous, $\forall z \in X$, we obtain that $A\tilde{U}(t)x = Au(t)$ belongs to $C([0,\infty);X)$.

As $Bu'(t) = Au(t)$, $\forall t \geq 0$, it follows that $(Bu')(\cdot)$ belongs to $C([0,\infty);X)$ too. ∎

Chapter V

Semigroups of class C_0 and the abstract Cauchy problem

Introduction

In this chapter we give the main definitions and some basic results for semigroups of linear continuous operators in a Banach space and then give a brief discussion of the abstract Cauchy problem which is associated with the given semigroup.

1. Let X be a Banach space and let $T(t)$ be a family of linear continuous operators, $X \to X$ depending on the parameter $t \geq 0$. We assume the following properties:

(a) $T(t)x \in C([0,\infty);X), \ \forall x \in X$

(b) $T(0)x = x, \ \forall x \in X$ (thus $T(0) = \text{Identity in } X$) (1.1)

(c) $T(t+s) = T(t)T(s), \ \forall t,s \in [0,\infty)$.

We say that we have a semigroup of class C_0.

Let us prove first:

Proposition 1.1 *For any C_0-semigroup the numerical-valued function*

$$t \to \|T(t)\| \tag{1.2}$$

is measurable and bounded on compact intervals of $[0,\infty)$.

Proof Consider, $\forall \alpha \in \mathbb{R}$, the set $\{t \in \mathbb{R}^+; \ \|T(t)\| > \alpha\} = \mathcal{E}_\alpha$. It is an open set in $\mathbb{R}^+$ (in fact take $t_0 \in \mathcal{E}_\alpha$; then $\|T(t_0)\| = \sup\limits_{\|x\|=1} \|T(t_0)x\| > \alpha$ and $\exists x_0 \in X, \ \|x_0\| = 1$, such that $\|T(t_0)x_0\| > \alpha$ too.

Now use condition (a) in (1.1) so that $\exists \delta > 0$, such that $\|T(t)x_0\| > \alpha$ for $t_0 - \delta < t < t_0 + \delta$ or for $0 \leq t < \delta$ if $t_0 = 0$. It follows that $\|T(t)\|_{\mathcal{L}(X)} = \sup\limits_{\|x\|=1} \|T(t)x_0\| > \alpha$

for $t_0-\delta<t<t_0+\delta$ which means that $(t_0-\delta,\ t_0+\delta)\subset\mathcal{E}_\alpha$; $\mathcal{E}_\alpha$ is open in $[0,\infty)$ hence the function $t\to\|T(t)\|$ is measurable on $[0,\infty)$.

Next take any compact interval $[\alpha,\beta]$ in $[0,\infty)$; again from (1.1)-a, we see that

$$\|T(t)x\|\le M(\alpha,\beta,x)\ , \quad \forall t\in[\alpha,\beta],\ \forall x\in X\ . \tag{1.3}$$

Now use the U.B.T. (Chapter II), and get a number $\overline{M}(\alpha,\beta)$ (independent of $x\in X$), such that

$$\|T(t)\|_{\mathcal{L}(X)}\le\overline{M}\ , \quad \forall t\in[\alpha,\beta]\ . \tag{1.4}$$

■

In order to establish our next result (an exponential bound for the function $t\to\|T(t)\|$) we shall give a preliminary result – also interesting in itself – about measurable subadditive functions.

Consider in fact measurable extended real valued functions $W(t)$ on $[0,\infty)$, with the following properties:

(i) $W(t)<+\infty,\ \ \forall t\ge 0$;

(ii) $W(t_1+t_2)\le W(t_1)+W(t_2),\ \ \forall t_1,t_2\in[0,\infty)$.

Then $W(t)$ is bounded above in any compact interval $[\alpha,\beta]$ of $[0,\infty]$ (see [5], p. 241).

One obtains also, if $w_0=\inf\limits_{t\ge 0}\dfrac{W(t)}{t}$, that $-\infty\le w_0<+\infty$ and $w_0=\lim\limits_{t\to\infty}\dfrac{W(t)}{t}$ (see for instance [2], p. 618, vol. I).

We now prove:

Proposition 1.2 *For each $w>w_0$, there is $M_w>0$, such that*

$$\|T(t)\|\le M_w e^{wt}\ , \quad \forall t\ge 0\ . \tag{1.5}$$

Proof Consider the function $W(t)=\log\|T(t)\|,\ 0\le t<\infty$. We have a measurable function, $-\infty\le W(t)<+\infty$ which is subadditive (as follows from (1.1)-c). From previous remarks we get

$$-\infty\le w_0=\lim_{t\to\infty}\frac{1}{t}\ \log\|T(t)\|<+\infty\ . \tag{1.6}$$

Accordingly, if $w > w_0$ we get $t_0 \in [0, \infty)$, such that

$$\frac{1}{t}\log\|T(t)\| < w \quad \text{for } t \geq t_0 \tag{1.7}$$

that is

$$\|T(t)\| \leq e^{wt}, \quad t \geq t_0 . \tag{1.8}$$

Put also $M_w^1 = \sup_{0<t\leq t_0} \|T(t)\|$, and we now see that

$$\|T(t)\| \leq \frac{M_w^1}{C_0} e^{wt}, \quad 0 \leq t \leq t_0$$

where $C_0 = \inf_{[0,t_0]} e^{wt}$, hence

$$\|T(t)\| \leq \max\left(\frac{M_w^1}{C_0}, 1\right) e^{wt}, \quad \forall t \geq 0 . \tag{1.9}$$

■

2. A fundamental concept in the theory of C_0 operator semigroups is that of 'infinitesimal generator'.

Given a C_0-semigroup $T(t)$ in the B-space X we consider the (strong) (right)-derivative for $t = 0$:

$$\lim_{\substack{h \to 0 \\ h > 0}} \frac{1}{h}[T(h)x - x], \quad \text{where } x \in X .$$

Actually, it so happens that in general this derivative will only exist for a (strict) subset of X. Then we define:

$$\mathcal{D} = \left\{ x \in X, \ \frac{d^+}{dt} T(t)x \bigg|_{t=0} \ \text{exists} \right\} \tag{2.1}$$

and accordingly, an operator A, $\mathcal{D} \to X$ is thus defined by

$$Ax = \lim_{\substack{\eta \to 0 \\ \eta > 0}} \frac{T(\eta)x - x}{\eta}, \quad \forall x \in \mathcal{D} . \tag{2.2}$$

It is readily seen that $\mathcal{D}$ is a linear subspace of X and then that A is a linear operator.

By definition, this operator A *is the infinitesimal generator of the semigroup* $T(t)$.

A simple but important result appears now as:

Proposition 2.1 *For any $x \in \mathcal{D} = \mathcal{D}(A)$ the strong derivative*

$$\frac{\mathrm{d}}{\mathrm{d}t} T(t)x$$

exists $\forall t \geq 0$ and one has equality

$$\frac{\mathrm{d}}{\mathrm{d}t} T(t)x = AT(t)x = T(t)Ax , \quad \forall x \in \mathcal{D}(A) . \tag{2.3}$$

In this statement we can 'implicitly' read:

(a) If $x \in \mathcal{D}(A)$ then $T(t)x \in \mathcal{D}(A)$, $\forall t \geq 0$ (we can say that $\mathcal{D}(A)$ is invariant under $T(t)$).

(b) The operators $T(t)$ and A commute on $\mathcal{D}(A)$.

(c) The derivative $\frac{\mathrm{d}}{\mathrm{d}t} T(t)x$, for $x \in D(A)$, is a continuous function on $[0, \infty)$.

(d) The function, $[0, \infty) \to X$, $u(t)$, defined by $u(t) = T(t)x$, $x \in D(A)$, is a solution to the 'abstract Cauchy problem':

$$u'(t) = Au(t) , \quad u(0) = x , \quad 0 \leq t < +\infty . \tag{2.4}$$

Wc shall pass now to the:

Proof of Proposition 2.1 Take first $\eta > 0$ and then write

$$\frac{1}{\eta}[T(t+\eta)x - T(t)x] = T(t)\frac{1}{\eta}[T(\eta)x - x] = \frac{T(\eta) - I}{\eta} T(t)x . \tag{2.5}$$

If $x \in \mathcal{D}(A)$ we find from the above that $\lim\limits_{\substack{\eta \to 0 \\ \eta > 0}} \frac{1}{\eta}[T(t+\eta)x - T(t)x]$ exists and equals $T(t) \cdot Ax$.

This also means that $\lim\limits_{\substack{\eta \to 0 \\ \eta > 0}} \frac{T(\eta)-I}{\eta} T(t)x$ exists, so that (for $x \in \mathcal{D}(A)$), $T(t)x \in \mathcal{D}(A)$ too, and

$$AT(t)x = T(t)Ax \ . \tag{2.6}$$

Thus, so far we got that, for $x \in \mathcal{D}(A)$, the right-derivative $\frac{\mathrm{d}^+}{\mathrm{d}t} T(t)x$ exists and equals

$$T(t)Ax = AT(t)x \ . \tag{2.7}$$

Looking now at the left-derivative, we consider the expression (for $x \in \mathcal{D}(A)$) $\frac{T(t-\eta)-T(t)}{-\eta} x$, $\eta > 0$, $t > 0$, η small enough.

This can be written as

$$\frac{T(t-\eta)-T(t-\eta+\eta)}{-\eta} x = T(t-\eta) \frac{T(\eta)-I}{\eta} x \ .$$

Now, we shall see that

$$\lim_{\substack{\eta \to 0 \\ \eta > 0}} T(t-\eta) \frac{T(\eta)-I}{\eta} x = T(t)Ax \ , \qquad \forall x \in \mathcal{D}(A) \ . \tag{2.8}$$

In fact, their difference is

$$T(t-\eta)\frac{T(\eta)-I}{\eta} x - T(t)Ax = T(t-\eta)\left[\frac{T(\eta)-I}{\eta} x - Ax\right] + T(t-\eta)Ax - T(t)Ax \ . \tag{2.9}$$

We next estimate, for $\eta > 0$, $\eta < t$:

$$\left\| T(t-\eta)\left[\frac{T(\eta)-I}{\eta} x - Ax\right] \right\|$$

$$\leq \|T(t-\eta)\| \left\| \frac{T(\eta)-I}{\eta} x - Ax \right\|$$

$$\leq C_t \left\| \frac{T(\eta)-I}{\eta} x - Ax \right\| \quad (\text{where } C_t = \sup_{0 \leq s \leq t} \|T(s)\|)$$

and then we readily get (2.8). Therefore $\frac{\mathrm{d}}{\mathrm{d}t} T(t)x$ exists, for $x \in \mathcal{D}(A)$, $t > 0$ and $\frac{\mathrm{d}}{\mathrm{d}t} T(t)x = T(t)Ax = AT(t)x$, which is (2.3). ■

Remark Looking at (2.4), it suggests that $u(t) = T(t)x = \mathrm{e}^{At}x = \mathrm{e}^{At}u(0)$. The semigroup $T(t)$ appears to be the abstract exponential function $\exp(At)$.

(This is also suggested by the 'functional equation' (1.1)-c which must be satisfied by any semigroup.)

Let us now explain the following:

Example 2.1 Consider the uniformly well-posed Cauchy problem (1.3)-(1.4), Chapter IV, and its associated semigroup $\widetilde{U}(t)$. Denote (for this example) by $\widetilde{A}$, the infinitesimal generator of this semigroup. We shall now see that

$$\mathcal{D}(A) \cap \mathcal{D}(B) \subset \mathcal{D}(\widetilde{A}) \quad \text{and} \quad B\widetilde{A}x = Ax, \quad \forall x \in \mathcal{D}(A) \cap \mathcal{D}(B). \tag{2.10}$$

In fact, let $x \in \mathcal{D}(A) \cap \mathcal{D}(B)$; then $\widetilde{U}(t)x = U(t)x = u(t)$ is a solution to the equation

$$Bu'(t) = Au(t) \quad \text{on } [0, \infty), \quad u(0) = x. \tag{2.11}$$

In particular we obtain that $u'_+(0)$ exists and

$$Bu'_+(0) = Au(0) = Ax. \tag{2.12}$$

On the other hand we see that $u'_+(0) = \frac{\mathrm{d}}{\mathrm{d}t} \widetilde{U}(t)x \Big|_{t=0} = \widetilde{A}x$ exists and accordingly $x \in \mathcal{D}(\widetilde{A})$.

Now (2.12) reads:

$$B\widetilde{A}x = Ax, \quad \forall x \in \mathcal{D}(A) \cap \mathcal{D}(B).$$

Remark If the inverse B^{-1} exists then (at least formally) $\tilde{A}x = B^{-1}Ax$ and equation (2.11) is also (formally at least) $u'(t) = \left(B^{-1}A\right)u(t)$. Then $\tilde{U}(t) = \exp\left(B^{-1}At\right)$, $t \geq 0$, which shows why the generator $\tilde{A}$ should be $B^{-1}A$ and then $B\tilde{A}$ should be equal to A.

3. In order to study further properties of C_0-semigroups and their generator we shall explain (briefly) how one extends to continuous functions from a real interval into a Banach space the classical concept of Riemann integral.

Let X be a B-space and $[a,b]$ a closed (bounded) interval on $\mathbb{R}$. A partition of $[a,b]$ is a finite ordered set of points $\pi = (t_0, t_1, \ldots, t_n)$ with $a = t_0 < t_1 < \ldots < t_n = b$. The norm of the partition π is the maximum length of the subinterval $[t_{i-1}, t_i]$, $i = 1,2,\ldots,n$ (denoted $\|\pi\|$).

Consider the 'Riemann sums':

$$S_{f,\pi} = \sum_{k=1}^{n} f(\xi_k)(t_k - t_{k-1}) \,,$$

where $\xi_k \in [t_{k-1}, t_k]$ and f is a bounded function $[a,b] \to X$.

The function f is said to be R-integrable if there exists an element $z \in X$ such that

$$\lim_{\|\pi\| \to 0} S_{f,\pi} = z \tag{3.1}$$

which means precisely the following:

$\forall \varepsilon > 0$, $\exists \delta(\varepsilon) > 0$ such that $\left\| S_{f,\pi} - z \right\|_X < \varepsilon$ if $\|\pi\| < \delta(\varepsilon)$ independently of the choice of ξ_k in $[t_{k-1}, t_k]$.

If this happens (with a unique element z) we denote

$$\int_a^b f = z = \int_a^b f(\sigma)\, \mathrm{d}\sigma \,. \tag{3.2}$$

We next prove

Proposition 3.1 *Let f be a bounded function,* $[a,b]\to X$, *and let us assume that* $\forall\varepsilon>0$, $\exists\delta(\varepsilon)>0$ *such that if* π_1,π_2 *are partitions of* $[a,b]$ *with* $\|\pi_1\|$, $\|\pi_2\|<\delta$, *then* $\left\|S_{f,\pi_1}-S_{f,\pi_2}\right\|<\varepsilon$ *(for any choice of* ξ_k,η_e*). Then f is R-integrable.*

Proof Consider a sequence of partitions (π_q) where $\|\pi_q\|<\frac{1}{q}$, $q=1,2,\ldots$. Let $z_q=\sum_{i=1}^{n_q} f(t_i^q)(t_i^q-t_{i-1}^q)$, Riemann sums associated to the partition $\pi_q=(t_0^q<t_1^q\ldots<t_n^q)$ and $\xi_{i,q}=t_{i,q}$.

Now, the sequence (z_q) is a Cauchy sequence in X (given $\varepsilon>0$ and $\bar{q}>\frac{1}{\delta(\varepsilon)}$ we get $\left\|z_q-z_r\right\|_X<\varepsilon$ for $q,r>\bar{q}$). Let $z=\lim_{q\to\infty} z_q$, and take any partition π of $[a,b]$. Take also a Riemann sum $\sum f(\xi_k)(t_k-t_{k-1})=S_{f,\pi}$ and write $S_{f,\pi}-z=S_{f,\pi}-S_{f,\pi_q}+S_{f,\pi_q}-z$ so that $\left\|S_{f,\pi}-z\right\|_X<\left\|S_{f,\pi}-S_{f,\pi_q}\right\|_X+\left\|S_{f,\pi_q}-z\right\|_X$. Given $\varepsilon>0$ take $\delta\left(\frac{\varepsilon}{2}\right)$, so that $\left\|S_{f,\pi}-S_{f,\pi_q}\right\|_X<\frac{\varepsilon}{2}$ for $\|\pi\|<\delta$, $\|\pi_q\|<\delta$ while $\left\|S_{f,\pi_q}-z\right\|_X<\frac{\varepsilon}{2}$ for $q\geq\bar{q}$. Thus, as indeed $\|\pi_q\|<\delta$ if q is sufficiently large, we obtain

$$\left\|S_{f,\pi}-z\right\|_X<\frac{\varepsilon}{2}+\frac{\varepsilon}{2}=\varepsilon \quad \text{if} \quad \|\pi\|<\delta .$$

Next follows the fundamental result in:

Proposition 3.2 *If f,* $[a,b]\to X$ *is a continuous function it is R-integrable.*

Proof As is well-known, the continuous functions $[a,b]\to X$ are bounded and also uniformly continuous. Thus, $\forall\varepsilon>0$, $\exists\delta(\varepsilon)>0$, such that, if $|t_1-t_2|<\delta$, $t_1,t_2\in[a,b]$, then $\left\|f(t_1)-f(t_2)\right\|_X<\varepsilon$.

Consider now two partitions of $[a,b]$, π_1, π_2, such that $\|\pi_1\|<\delta$, $\|\pi_2\|<\delta$. Suppose that $\pi_1=(t_0,t_1,\ldots,t_n)$ and let π' be a partition which includes all points of π_1 and π_2: $\pi'=(s_0,s_1,\ldots,s_m)$.

Consider one summand of $S_{f,\pi_1}=\sum f(\xi_k)(t_k-t_{k-1})$, that is $f(\xi_i)(t_i-t_{i-1})$; we may assume that

$$t_{i-1}=s_{j-1}<s_j\ldots<s_k=t_i . \tag{3.3}$$

We obtain (with $\eta_e \in [s_{e-1}, s_e]$)

$$f(\xi_i)(t_i - t_{i-1}) - \sum_{l=j}^{k} f(\eta_e)(s_e - s_{e-1}) = \sum_{l=j}^{k} (f(\xi_i) - f(\eta_e))(s_e - s_{e-1})$$

where $\|f(\xi_i) - f(\eta_e)\| < \varepsilon$ and accordingly

$$\left\| f(\xi_i)(t_i - t_{i-1}) - \sum_{l=j}^{k} f(\eta_e)(s_e - s_{e-1}) \right\|_X < \varepsilon(t_i - t_{i-1}) . \tag{3.4}$$

Consequently one obtains:

$$\left\| S_{f,\pi_1} - S_{f,\pi'} \right\| < \varepsilon(b-a) . \tag{3.5}$$

In a similar way

$$\left\| S_{f,\pi_2} - S_{f,\pi'} \right\| < \varepsilon(b-a) , \tag{3.6}$$

hence

$$\left\| S_{f,\pi_1} - S_{f,\pi_2} \right\| < 2\varepsilon(b-a) \tag{3.7}$$

and then apply Proposition 3.1. ■

We have therefore the existence result of the Riemann integral for X-valued continuous functions. Other properties of this integral are 'easier' to prove so that we accept them without further ado.

We are therefore ready for:

Proposition 3.3 *If $T(t)$ is a C_0-semigroup in the B-space X, the domain $\mathcal{D} = \mathcal{D}(A)$ of its infinitesimal generator is dense in X.*

Proof First we show that, $\forall x \in X$, $\forall t \geq 0$, the element of X: $\int_0^t T(\sigma)x \, d\sigma$ belongs to $\mathcal{D}(A)$.

Note - to begin with - that the R-integral commutes with linear continuous operators on X - which means that, if $f(\cdot) \in C([a,b];X)$ and $M \in \mathcal{L}(X)$ then $(Mf)(\cdot) \in C([a,b];X)$ and

$$M\int_a^b f(\sigma)\,\mathrm{d}\sigma = \int_a^b (Mf)(\sigma)\,\mathrm{d}\sigma \tag{3.8}$$

(this is readily established taking Riemann sums convergent to the integral).

Consider now the expression, for $h>0,\ x \in X$,

$$\frac{T(h)-I}{h}\int_0^t T(\sigma)x\,\mathrm{d}\sigma \tag{3.9}$$

which is also written as

$$\begin{aligned}
&\frac{1}{h}\int_0^t [T(\sigma+h)x - T(\sigma)x]\,\mathrm{d}\sigma \\
&\quad = \frac{1}{h}\int_h^{t+h} T(\sigma)x\,\mathrm{d}\sigma - \frac{1}{h}\int_0^t T(\sigma)x\,\mathrm{d}\sigma \\
&\quad = -\frac{1}{h}\int_0^h T(\sigma)x\,\mathrm{d}\sigma + \frac{1}{h}\int_t^{t+h} T(\sigma)x\,\mathrm{d}\sigma\ .
\end{aligned} \tag{3.10}$$

Accordingly we see that

$$\lim_{\substack{h\to 0\\ h>0}} \frac{T(h)-I}{h}\int_0^t T(\sigma)x\,\mathrm{d}\sigma$$

exists and equals $T(t)x - x$ and also that

$$A\int_0^t T(\sigma)x\,\mathrm{d}\sigma = T(t)x - x\ . \tag{3.11}$$

At this stage, the density of $\mathcal{D}(A)$ in X is a direct consequence of the following: $\forall x \in X$, the sequence (x_n) where $x_n = n\int_0^{\frac{1}{n}} T(\sigma)x \, \mathrm{d}\sigma$ is convergent to x.

(In fact, $x_n - x = n\int_0^{\frac{1}{n}} [T(\sigma)x - x] \, \mathrm{d}\sigma$, hence

$$\|x_n - x\| \le n\int_0^{\frac{1}{n}} \|T(\sigma)x - x\| \, \mathrm{d}\sigma \le \sup_{0\le\sigma\le\frac{1}{n}} \|T(\sigma)x - x\| \to 0 \quad \text{as} \quad n \to \infty . \qquad \blacksquare$$

Another important fact in the theory of C_0-semigroups is expressed in:

Proposition 3.4 *The infinitesimal generator of a C_0-semigroup is a closed operator.*

Proof First we note the following property. If $T(t)$ is a C_0-semigroup with generator A and if $x \in \mathcal{D}(A)$ then

$$A\int_0^t T(\sigma)x \, \mathrm{d}\sigma = \int_0^t T(\sigma)Ax \, \mathrm{d}\sigma . \tag{3.12}$$

In fact, using 2.3, we get

$$\int_0^t T(\sigma)Ax \, \mathrm{d}\sigma = \int_0^t \frac{\mathrm{d}}{\mathrm{d}\sigma} T(\sigma)x \, \mathrm{d}\sigma = T(t)x - T(0)x = A\int_0^t T(\sigma)x \, \mathrm{d}\sigma \qquad \text{(by 3.11)}.$$

Consider now, as usual, a sequence (x_n) in $\mathcal{D}(A)$ such that $x_n \to x_0$ and $Ax_n \to y_0$. We have, by (3.11)-(3.12), $\forall n \in \mathbb{N}$, the equality

$$T(t)x_n - x_n = \int_0^t T(\sigma)Ax_n \, \mathrm{d}\sigma$$

which becomes, for $n \to \infty$,

$$T(t)x_0 - x_0 = \int_0^t T(\sigma)y_0 \, d\sigma \quad \text{and} \quad \frac{1}{t}\left[T(t)x_0 - x_0\right] = \frac{1}{t}\int_0^t T(\sigma)y_0 \, d\sigma \; .$$

Therefore $\lim\limits_{\substack{t \to 0 \\ t>0}} \frac{1}{t}\left[T(t)x_0 - x_0\right]$ exists and equals y_0 so that $x_0 \in \mathcal{D}(A)$ and $Ax_0 = y_0$.

This proves the closedness of the generator A.

4. In the final section of this chapter we consider the 'abstract Cauchy problem' for a first-order equation

$$u'(t) = Au(t) + f(t) \tag{4.1}$$

with 'right-hand side' f and a closed linear operator with dense domain: A, $\mathcal{D}(A) \subset X \to X$, where X is as previously, a Banach space over the real or complex field.

Precisely, f is defined on $[0, \infty)$ with range in X; one looks for a function $u(\cdot)$, $[0, \infty) \to \mathcal{D}(A)$, X-strongly differentiable, that is $u'(t)$ exists in X-strong, $\forall t \geq 0$ and such that (4.1) holds, $\forall t \geq 0$.

If we ask for a solution $u(\cdot)$ of (4.1) satisfying the supplementary condition

$$u(0) = u_0 \quad \text{a given element in } \mathcal{D}(A) \tag{4.2}$$

we have the abstract Cauchy problem (already encountered in (2.4) - for $f \equiv \mathbf{0}$).

We establish the following:

Theorem 4.1 *Let us assume that A is an infinitesimal generator of a C_0-semigroup $T(t)$. Then*

(a) *If $u(t)$ is an X-differentiable solution of the equations*

$$u'(t) = Au(t) \, , \quad 0 \leq t < \infty, \; u(0) = \mathbf{0} \tag{4.3}$$

it follows that $u(t) = \mathbf{0}$, $\forall t > 0$.

(b) *If $f(\cdot) \in C^1([0, \infty); X)$ and $u_0 \in \mathcal{D}(A)$, the function $u(\cdot)$ represented by the formula*

$$u(t) = T(t)u_0 + \int_0^t T(t-s)f(s)\,\mathrm{d}s\,, \qquad t \ge 0 \tag{4.4}$$

is the solution to the (abstract) Cauchy problem (4.1)-(4.2).

Proof We start with a demonstration of the uniqueness part (a).

Consider, for $0 \le s \le t$, the function $v_t(s)$ which is defined by the formula

$$v_t(s) = T(t-s)u(s)\,, \qquad 0 \le s \le t\,. \tag{4.5}$$

Next, consider the situation $0 < s < t$ and show that $v_t'(s)$ exists and equals $\mathbf{0}$. The differential quotient

$$\frac{1}{\tau}\left[v_t(s+\tau) - v_t(s)\right] \quad \text{(for small } \tau\text{)}$$

can be written as

$$\begin{aligned}\frac{1}{\tau}&\left[v_t(s+\tau) - v_t(s)\right]\\ &= \frac{1}{\tau}[T(t-s-\tau)u(s+\tau) - T(t-s)u(s)] \qquad (4.6)\\ &= \frac{1}{\tau}T(t-s-\tau)[u(s+\tau) - u(s)] + \frac{1}{\tau}[T(t-s-\tau) - T(t-s)]u(s)\,.\end{aligned}$$

Next, assume $\tau > 0$ (and small) and write (4.6) in the form

$$\frac{1}{\tau}\left[v_t(s+\tau) - v_t(s)\right] - T(t-s-\tau)\frac{1}{\tau}[u(s+\tau) - u(s)] - T(t-s-\tau)\frac{1}{\tau}[T(\tau) - I]u(s) \tag{4.7}$$

while, afterwards, for $\tau < 0$ (and small) we write (4.6) in the form

$$\frac{1}{\tau}\left[v_t(s+\tau) - v_t(s)\right] = T(t-s-\tau)\left[\frac{u(s+\tau) - u(s)}{\tau}\right] - T(t-s)\left[\frac{T(-\tau) - I}{-\tau}\right]u(s)\,. \tag{4.8}$$

Let us establish also the limit

$$\lim_{\tau\to 0} T(t-s-\tau)\frac{1}{\tau}[u(s+\tau)-u(s)] = T(t-s)u'(s) \ . \tag{4.9}$$

Taking difference one gets

$$T(t-s-\tau)\frac{1}{\tau}[u(s+\tau)-u(s)] - T(t-s)u'(s)$$
$$= T(t-s-\tau)\left\{\frac{1}{\tau}[u(s+\tau)-u(s)] - u'(s)\right\} + T(t-s-\tau)u'(s) - T(t-s)u'(s) \ . \tag{4.10}$$

Strong continuity of the semigroup $T(t)$ (condition (a) in (1.1)) then implies that

$$\lim_{\tau\to 0} T(t-s-\tau)u'(s) = T(t-s)u'(s) \ .$$

On the other hand, the first term in the right-hand side of (4.10) is estimated (using (1.5)) by

$$\left\| T(t-s-\tau)\left\{\frac{1}{\tau}[u(s+\tau)-u(s)] - u'(s)\right\}\right\| \le Me^{w(t-s-\tau)}\left\|\frac{1}{\tau}[u(s+\tau)-u(s)] - u'(s)\right\| \tag{4.11}$$

and (4.11) readily implies that

$$\lim_{\tau\to 0}\left\| T(t-s-\tau)\left\{\frac{1}{\tau}[u(s+\tau)-u(s)] - u'(s)\right\}\right\| = 0 \ .$$

Thus, (4.9) is true.

In a quite similar way we can see that

$$\lim_{\substack{\tau\to 0\\ \tau>0}} T(t-s-\tau)\frac{1}{\tau}[T(\tau)-I]u(s) = \lim_{\substack{\tau\to 0\\ \tau<0}} T(t-s)\left[\frac{1}{-\tau}(T(-\tau)-I)\right]u(s) = T(t-s)Au(s) \tag{4.12}$$

(note that $u(\cdot)$ is a solution of $u' = Au$ means, in particular, that $u(s) \in \mathcal{D}(A)$, $\forall s \ge 0$).

From all the above simple reasonings we get in fact, for $0 < s < t$, that $v_t'(s)$ exists and equals

$$T(t-s)u'(s) - T(t-s)Au(s) = T(t-s)[u'(s) - Au(s)] = \mathbf{0} \ . \tag{4.13}$$

Now, (4.13) implies (see the lemma below), that, if $0 < \varepsilon < s < t$, then $v_t(s) = v_t(\varepsilon)$, that is

$$T(t-s)u(s) = T(t-\varepsilon)u(\varepsilon)\,, \qquad 0<\varepsilon<s<t\,. \tag{4.14}$$

Next, note that $\lim\limits_{\substack{\varepsilon\to 0\\ \varepsilon>0}} T(t-\varepsilon)u(\varepsilon) = T(t)u(0) = \mathbf{0}$.

Accordingly, the left-hand side in (4.14), which is independent of $\varepsilon>0$, must be $\mathbf{0}$. Thus, we now have

$$T(t-s)u(s) = \mathbf{0} \quad \text{for} \quad 0<s<t\,. \tag{4.15}$$

Then $\lim\limits_{\substack{t\to s\\ t>s}} T(t-s)u(s) = u(s) = \mathbf{0}$ too, $\forall s>0$ (conditions (a)-(b) in (1.1)). ■

Remark In the proof above we used the following very useful result.

Lemma *If the function* $f(\cdot)$, $a<t<b\to X$, *has zero (strong) derivative:* $f'(t)=\mathbf{0}$, $a<t<b$, *then* $f(t)=f(t_0)$, $\forall t\in(a,b)$, *where* t_0 *is fixed in* (a,b).

Proof Let X' be the dual space to X; let us first assume that the field of X is $\mathbb{R}$. Then, $\forall x'\in X'$, the real-valued function $x'(f(t))$, $a<t<b$, has zero-derivative. Accordingly $x'(f(t)) = x'(f(t_0))$, $\forall t\in(a,b)$, $\forall x'\in X'$. Use of the Note to Corollary 1.1 in Chapter III, shows that $f(t)=f(t_0)$, $\forall t\in(a,b)$.

If X is a vector space over the complex field $\mathbb{C}$, we make similar deductions on the functions $\operatorname{Re} x'(f(t))$, $\operatorname{Im} x'(f(t))$. ■

We next continue with a proof of part (b) (existence of the solution for the abstract Cauchy problem).

We know that, as $u_0\in\mathcal{D}(A)$, the function $t\to T(t)u_0 = v(t)$, is a solution on $[0,\infty)$ of the equation $v'(t) = Av(t)$, $v(0)=u_0$. Thus, it is sufficient to prove that the function $t\to\int_0^t T(t-s)f(s)\,\mathrm{d}s = w(t)$ is a solution on $[0,\infty)$ of the equation

$$w'(t) = Aw(t) + f(t)\,, \qquad w(0)=\mathbf{0}\,, \qquad \text{and} \qquad w'(\cdot)\in C([0,\infty);X)\,.$$

In order to do this, write first $w(\cdot)$ in the form

$$w(t)=\int_0^t T(\sigma)f(t-\sigma)\,\mathrm{d}\sigma\ . \tag{4.16}$$

Then we get, for $h>0$,

$$w(t+h)=\int_0^{t+h} T(\sigma)f(t+h-\sigma)\,\mathrm{d}\sigma=\int_0^t T(\sigma)f(t+h-\sigma)\,\mathrm{d}\sigma$$

$$+\int_t^{t+h} T(\sigma)f(t+h-\sigma)\,\mathrm{d}\sigma\ .$$

Therefore, the differential quotient $\frac{1}{h}[w(t+h)-w(t)]$ is given by the expression

$$\int_0^t T(\sigma)\frac{1}{h}[f(t+h+\sigma)-f(t-\sigma)]\,\mathrm{d}\sigma+\frac{1}{h}\int_t^{t+h} T(\sigma)f(t+h-\sigma)\,\mathrm{d}\sigma\ . \tag{4.17}$$

Now, we can see that

$$\lim_{\substack{h\to 0\\ h>0}}\int_0^t T(\sigma)\frac{1}{h}[f(t+h-\sigma)-f(t-\sigma)]\,\mathrm{d}\sigma=\int_0^t T(\sigma)f'(t-\sigma)\,\mathrm{d}\sigma$$

and

$$\lim_{\substack{h\to 0\\ h>0}}\frac{1}{h}\int_t^{t+h} T(\sigma)f(t+h-\sigma)\,\mathrm{d}\sigma=T(t)f(0)\ .$$

We have in fact

$$\left\|\int_0^t T(\sigma)\left\{\frac{1}{h}[f(t+h-\sigma)-f(t-\sigma)]-f'(t-\sigma)\right\}\mathrm{d}\sigma\right\|$$

$$\leq \sup_{0\leq\sigma\leq t}\|T(\sigma)\|\cdot\sup_{0\leq\sigma\leq t}\left\|\frac{1}{h}[f(t+h-\sigma)-f(t-\sigma)]-f'(t-\sigma)\right\|\cdot t \tag{4.18}$$

and also

$$\frac{1}{h}\int_t^{t+h} T(\sigma)f(t+h-\sigma)\,\mathrm{d}\sigma-T(t)f(0)=\frac{1}{h}\int_t^{t+h}[T(\sigma)f(t+h-\sigma)-T(t)f(0)]\,\mathrm{d}\sigma$$

$$=\frac{1}{h}\int_t^{t+h} T(\sigma)[f(t+h-\sigma)-f(0)]\,\mathrm{d}\sigma+\frac{1}{h}\int_t^{t+h}[T(\sigma)f(0)-T(t)f(0)]\,\mathrm{d}\sigma\ . \tag{4.19}$$

Next, note that

$$\frac{1}{h}[f(t+h-\sigma)-f(t-\sigma)]=\frac{1}{h}\int_t^{t+h} f'(\tau-\sigma)\,\mathrm{d}\tau$$

and therefore

$$\frac{1}{h}[f(t+h-\sigma)-f(t-\sigma)]-f'(t-\sigma)=\frac{1}{h}\int_t^{t+h}[f'(\tau-\sigma)-f'(t-\sigma)]\,\mathrm{d}\tau$$

whence we have the estimate

$$\left\|\frac{1}{h}[f(t+h-\sigma)-f(t-\sigma)]-f'(t-\sigma)\right\|\leq \sup_{t\leq\tau\leq t+h}\|f'(\tau-\sigma)-f'(t-\sigma)\|$$

which tends to 0 as $h\to 0$, uniformly for $0\leq\sigma\leq t$ (use uniform continuity of f').

We also estimate right-hand side terms in (4.19) and obtain

$$\left\|\frac{1}{h}\int_t^{t+h} T(\sigma)[f(t+h-\sigma)-f(0)]\,\mathrm{d}\sigma\right\|\leq \sup_{t\leq\sigma\leq t+h}\|T(\sigma)\|\cdot\|f(t+h-\sigma)-f(0)\|\,.$$

Here $\sup_{t\leq\sigma\leq t+h}\|f(t+h-\sigma)-f(0)\|\to 0$ as $h\to 0$, due to the continuity of $f(\cdot)$ for $t=0$.

Also

$$\left\|\frac{1}{h}\int_t^{t+h}[T(\sigma)f(0)-T(t)f(0)]\,\mathrm{d}\sigma\right\|\leq \sup_{t\leq\sigma\leq t+h}\|T(\sigma)f(0)-T(t)f(0)\|$$

which tends to 0 as the function $\sigma\to T(\sigma)f(0)$ is continuous for $\sigma=t$.

From the above computations and estimates we derive: the right-derivative of function $w(t)$ exists, $\forall t\geq 0$, and is given by the expression

$$w'_+(t)=T(t)f(0)+\int_0^t T(\sigma)f'(t-\sigma)\,\mathrm{d}\sigma \tag{4.20}$$

which also indicates that $w'_+(t)\in C([0,\infty);X)$.

If we now use a well-known result (see for instance Lemma 5.1, p. 25 in [17]) we obtain that the derivative $w'(t)$ exists, $\forall t>0$, and is given by (4.20).

The last step in the proof will consist in the following: $w(t)\in\mathcal{D}(A)$, $\forall t\geq 0$, and $w'(t)=Aw(t)+f(t)$, $\forall t\geq 0$.

We again write $w(t)=\int_0^t T(t-s)f(s)\,\mathrm{d}s$; the differential quotient $\frac{1}{h}[w(t+h)-w(t)]$ (for $h>0$) is given by the expression

$$\begin{aligned}\frac{1}{h}\left\{\int_0^{t+h} T(t+h-s)f(s)\,\mathrm{d}s-\int_0^t T(t-s)f(s)\,\mathrm{d}s\right\}&=\int_0^t \frac{1}{h}[T(h)-I]T(t-s)f(s)\,\mathrm{d}s\\ +\frac{1}{h}\int_t^{t+h} T(t+h-s)f(s)\,\mathrm{d}s&=\frac{1}{h}[T(h)-I]\int_0^t T(t-s)f(s)\,\mathrm{d}s\\ &+\frac{1}{h}\int_t^{t+h} T(t+h-s)f(s)\,\mathrm{d}s\end{aligned} \tag{4.21}$$

(we used here the commutativity of bounded linear operators with Riemann integrals - see (3.8)).

Now, it is not difficult to see that

$$\lim_{h\to 0}\frac{1}{h}\int_t^{t+h} T(t+h-s)f(s)\,\mathrm{d}s$$

exists and equals $f(t)$. Also, we previously proved the existence of the limit

$$\lim_{h\to 0}\frac{w(t+h)-w(t)}{h}.$$

Then, from (4.21) we infer the existence of the limit

$$\lim_{\substack{h\to 0\\ h>0}}\frac{1}{h}[T(h)-I]\int_0^t T(t-s)f(s)\,\mathrm{d}s \tag{4.22}$$

which equals $-f(t)+w'(t)$.

Finally, we use the definition of the infinitesimal generator (see (2.2) above), to find that:

$$w(t)=\int_0^t T(t-s)f(s)\,\mathrm{d}s\in\mathcal{D}(A) \quad\text{and}\quad Aw(t)=-f(t)+w'(t) \tag{4.23}$$

which is (4.1), the equation which we were trying to solve. ■

Chapter VI
Compact linear operators

Introduction

This subclass of all linear continuous operators is broad enough to include many operators of interest and still narrow enough to allow the proof of significant theorems. Compact operators include many integral operators and yet have properties somewhat close to operators in finite-dimensional spaces.

1. One of the 'simplest' definitions of compact (linear) operators is as follows.

Definition 1.1 *Let E,F be normed spaces and let T, $E \to F$, be linear. Then, T is also compact if it has the following property: $\forall$ bounded sequence $(x_n)_1^\infty$ in E, the sequence $(Tx_n)_1^\infty$ has a convergent subsequence in F.*

The following result is not difficult to prove.

Proposition 1.1 *Any compact operator T is a continuous operator.* In fact, let us use Theorem 3.1, Chapter 1; thus, if T is not continuous, $E \to F$, there is a bounded set A in E, such that $T(A)$ is not bounded in F; accordingly, $\forall n \in \mathbb{N}$, there is $x_n \in A$, such that $\|Tx_n\|_F \geq n$. Obviously, there is no subsequence of $(Tx_n)_1^\infty$ which is convergent. ■

An important instance of compact operators are those continuous operators which are 'of finite rank': this means that $T(E)$ is a finite dimensional subspace of F. Now, if $(x_n)_1^\infty$ is a bounded sequence in E, $(Tx_n)_1^\infty$ will be a bounded sequence in $F_1 = T(E)$. As, loosely speaking, in finite dimensional spaces every bounded sequence has a convergent subsequence, we get our assertion.

Example 1 In a (pre) Hilbert space V, take elements $a_1, a_2, \ldots, a_n$, $b_1, b_2, \ldots, b_n$. Then, $\forall h \in V$, put $Th = \sum_{i=1}^{n} (h, a_i) b_i$.

This is obviously a linear operator, $V \to V$, and $\|Th\| \le \sum_{i>1}^{n} |(h, a_i)| \, \|b_i\| \le \|h\| \sum_{i=1}^{n} \|a_i\| \, \|b_i\|$, so that T is continuous. Finally, T maps V into the linear space spanned by elements $b_1, b_2, \ldots, b_n$, and this is a finite dimensional space.

Example 2 Consider the space $E = C[a,b]$ of all continuous functions $\varphi(\cdot)$, $[a,b] \to \mathbb{R}$, with the sup-norm. Take a function $\mathcal{K}(x,y)$, $[a,b] \times [a,b] \to \mathbb{R}$ which is continuous. Consider the mapping K, defined on E by:

$$(K\varphi)(x) = \int_a^b \mathcal{K}(x,y)\varphi(y) \,\mathrm{d}y \,. \tag{1.1}$$

Due to the uniform continuity of the function $\mathcal{K}(x,y)$ on the square $[a,b] \times [a,b]$, we see immediately that $(K\varphi)(\cdot) \in E$. Thus the operator K is linear continuous, $E \to E$ as follows from the estimate ($\forall x \in [a,b]$)

$$\begin{aligned} |(K\varphi)(x)| &\le \int_a^b |\mathcal{K}(x,y)| \, |\varphi(y)| \,\mathrm{d}y \\ &\le \|\varphi(\cdot)\|_{C[a,b]} \int_a^b |\mathcal{K}(x,y)| \,\mathrm{d}y \\ &\le \|\varphi(\cdot)\|_{C[a,b]} \sup_{[a,b]\times[a,b]} |\mathcal{K}(x,y)| \cdot (b-a) \,. \end{aligned} \tag{1.2}$$

The main point, however, in this example, is to show why K is a compact operator. Take therefore a sequence $(\varphi_p(\cdot))$ in E, such that

$$\sup_{x \in [a,b]} |\varphi_p(x)| = C \,, \qquad \forall p = 1,2,\ldots \,. \tag{1.3}$$

The image sequence under K is then $(K\varphi_p)(x) = \int_a^b \mathcal{K}(x,y)\varphi_p(y) \,\mathrm{d}y$.

This is an equibounded sequence of continuous functions; it is also an equicontinuous sequence as can be deduced from estimate

$$\left|K\varphi_p(x_1)-K\varphi_p(x_2)\right| \le \int_a^b \left|\mathcal{K}(x_1,y)-\mathcal{K}(x_2,y)\right|\left|\varphi_p(y)\right| \mathrm{d}y \ . \tag{1.4}$$

Therefore, the sequence $\left(K\varphi_p(\cdot)\right)_1^\infty$ satisfies the conditions of Arzela-Ascoli theorem, and one can extract a convergent (in $C[a,b]$) subsequence.

Now, an important first result concerning compact operators can be stated as:

Theorem 1.1 *Let E,F be Banach spaces. The set of compact operators in $\mathcal{L}(E,F)$ is closed in $\mathcal{L}(E,F)$ with respect to the operator norm.*

For the proof, using the present definition of compact operators, see [14], p. 91.

A further simple result concerns sums of compact operators:

Let E,F be vector normed spaces and T_1,T_2 be (linear) compact operators, $E \to F$. Then, operator T_1+T_2, $E \to F$ is also compact.

Proof Take a bounded sequence $(x_n)_1^\infty$ in E. There exists a subsequence $\left(x_{n_p}\right)_1^\infty$ such that $\lim\limits_{p\to\infty} T_1x_{n_p} = z_1$ (say) exists in F. Then, there exists also a further subsequence $\left(x_{n_{p_g}}\right)_1^\infty$ such that $\lim\limits_{g\to\infty} T_2x_{n_{p_g}} = z_2$ (say) exists also in F.

Thus, $\lim\limits_{g\to\infty} T_1x_{n_{p_g}} = z_1$ and $\lim\limits_{g\to\infty} T_2x_{n_{p_g}} = z_2$; accordingly $\lim\limits_{g\to\infty}(T_1+T_2)x_{n_{p_g}}$ exists and equals $z_1+z_2 \in F$. ■

There is also a result concerning products (right or left) of compact operators with continuous (linear) operators.

Let E be a normed vector space and T a compact operator, $E \to E$. Assume also that $A \in \mathcal{L}(E)$. Then the operators AT and TA are both compact, $E \to E$.

Take a bounded sequence (x_n) in E; there exists a subsequence $\left(x_{n_p}\right)$ such that $\lim\limits_{p\to\infty} Tx_{n_p} = z$, and then $\lim\limits_{p\to\infty} A\left(Tx_{n_p}\right) = Az$. The operator AT is compact. Next, if (x_n) is bounded in E, the sequence (Ax_n) is also bounded in E; a subsequence $\left(Ax_{n_p}\right)$ must exist, such that $\lim\limits_{p\to\infty} T\left(Ax_{n_p}\right)=z$. However, $T\left(Ax_{n_p}\right)=(TA)\left(x_{n_p}\right)$; thus TA is compact.

Consider now a Hilbert space H, a compact operator T, $H\to H$ and its adjoint operator T^* (given by the equality $(Th,k)=(h,T^*k)$, $\forall h,k\in H$. *Then* T^* *is also a compact operator.*

Proof Let $(h_n)_1^\infty$ be a bounded sequence in H. From the above, the operator TT^* is compact, $H\to H$. We find a subsequence $\left(h_{n_p}\right)$ such that $\lim\limits_{p\to\infty} TT^*h_{n_p}=z$.

Now, the sequence $\left(T^*h_{n_p}\right)$ is a Cauchy sequence in H:

$$\begin{aligned}\left\|T^*\left(h_{n_p}-h_{n_q}\right)\right\|^2 &= \left(h_{n_p}-h_{n_q},\ TT^*\left(h_{n_p}-h_{n_q}\right)\right)\\ &\le 2\sup_{p\in\mathbf{N}}\left\|h_{n_p}\right\|\left\|TT^*h_{n_p}-TT^*h_{n_q}\right\|\\ &<\varepsilon \quad \text{for } p,q\ge\bar{p}(\varepsilon)\ .\end{aligned}$$

■

We now terminate this section with the following:

Theorem 1.2 *Let T be a compact operator in the Hilbert space H and* $\lambda\in\mathbb{C}$, $\lambda\ne0$. *Let* $H_1=\{h\in H,\ (\lambda-T)h=\mathbf{0}\}$. *Then* H_1 *is a subspace of H of finite dimension.*

Proof As $T\in\mathcal{L}(H)$ we see that H_1 is a closed linear subspace of H. Take any sequence (h_n) such that $\|h_n\|=1$, $h_n\in H_1$. Then $\lambda h_n=Th_n$ and $h_n=\frac{1}{\lambda}Th_n$; as T is compact, $\exists$ a subsequence $\left(h_{n_p}\right)$ such that $\left(Th_{n_p}\right)$ is convergent; then $\left(h_{n_p}\right)=\left(\frac{1}{\lambda}Th_{n_p}\right)$ is also a convergent subsequence. This property of sequences on the unit ball of H_1 implies that H_1 is of finite dimension. ■

(A Hilbert space has finite dimension if there is a finite set of linearly independent elements in the space which spans the whole space; it can be seen that if a Hilbert space is *not* of finite dimension, then we can find in H an infinite sequence of linearly independent elements and even a sequence $\left(e_g\right)_1^\infty$ of mutually orthogonal elements of unit norm. But for such a sequence the mutual distance $\|e_g - e_k\|_H$, $g \neq k$, is constant $= \sqrt{2}$ which precludes the existence of any convergent subsequence.)

2. An interesting (simple) property of compact operators in a Banach space is that they transform weakly convergent sequences in the (strongly) convergent ones.

Precisely, we have

Theorem 2.1 *Let X be a Banach space, and T, $X \to X$, be a linear compact operator. Let (x_n) a sequence in X, such that, $\forall F \in X'$ (the dual space to X), we have $F(x_n) \to F(x_0)$ with some element $x_0 \in X$.*

Then, $\lim_{n\to\infty} \|Tx_n - Tx_0\| = 0.$

Proof For all $F \in X'$, the numerical sequence $(F(x_n))$ is convergent, hence bounded. Accordingly, the set $\{x_n\}$ in X is weakly bounded, hence also bounded (Example 3, Chapter III, Section 2). Thus, the sequence (x_n) has a subsequence (x_{n_p}), such that: $\lim_{p\to\infty} Tx_{n_p} = z_0 \in X$.

Next, let us prove that, $\forall F \in X'$,

$$\lim_{n\to\infty} F(Tx_n) = F(Tx_0) . \tag{2.1}$$

Consider the dual operator to T; T' acting in X' (see Example 1, Chapter III, Section 15); we know that $F(Tx) = (T'F)(x)$, $\forall x \in X$, $\forall F \in X'$, and also that $T' \in \mathcal{L}(X')$, $\|T'\| \leq \|T\|$. From (2.1) we get $\lim_{n\to\infty} (T'F)(x_n) = (T'F)(x_0)$ where $T'F = G \in X'$, and this gives (2.1).

Thus we now have the following situation:

$Tx_n \to Tx_0$ weakly and a subsequence (Tx_{n_p}) converges (in norm) to z_0.

We show that this gives $Tx_n \to Tx_0$ in norm.

This last assertion means that $\forall \varepsilon > 0$, $\exists \overline{n}(\varepsilon)$, such that $\|Tx_n - Tx_0\| < \varepsilon$, $\forall n \geq \overline{n}(\varepsilon)$. Therefore, if (Tx_n) does not converge to Tx_0, it follows that, $\exists \varepsilon_0 > 0$, such that $\forall n \in \mathbb{N}$, $\exists m > n$ and $\|Tx_m - Tx_0\| \geq \varepsilon_0$. Take now $n = 1$ and $m_1 > 1$ such that $\|Tx_{m_1} - Tx_0\| \geq \varepsilon$. Next, take $m_2 > m_1$, such that $\|Tx_{m_2} - Tx_0\| \geq \varepsilon$; we continue in the same way and obtain a subsequence (Tx_{m_k}), where $m_1 < m_2 < m_3 \ldots$, and $\|Tx_{m_k} - Tx_0\| \geq \varepsilon$.

Actually, $\left(x_{m_k}\right)$ is a bounded sequence, hence for a certain subsequence $\left(x_{m_{k_g}}\right)$ we get $\lim_{g\to\infty} Tx_{m_{k_g}} = z_1 \in X$. Also, from the above lower bound we get that $\|z_1 - Tx_0\| \ge \varepsilon$, $z_1 \ne Tx_0$.

Note, on the other hand, that $Tx_{m_{k_g}} \to Tx_0$ weakly. Also, we see obviously that $\lim_{g\to\infty} Tx_{m_{k_g}} = z_1$ weakly ('strong convergence implies weak convergence').

All this means that the sequence $\left(Tx_{m_{k_g}}\right)$ has two (different) weak limits, which is impossible (see Example, Chapter III, Section 1). ■

Chapter VII

Continuous symmetric operators in Hilbert spaces

Introduction

A fundamental class of linear operators in Hilbert space is composed of (everywhere defined) symmetric operators; they are necessarily continuous and enjoy many important and peculiar properties.

1. Thus, if H is a Hilbert space and A, $H \to H$, is a linear operator such that $(Ah,k)=(h,Ak)$, $\forall h,k \in H$, then it is also a continuous operator (see Chapter II). One typical instance of a symmetric operator is the so-called *projection operator*.

If H is a Hilbert space and M is a closed linear subspace, then, as seen in the Projection Theorem (Theorem 2.1, Chapter I), every $x \in H$ can be written in one and only one way in the form $x = y+z$, where $y \in M$, $z \in M^{\perp}$. The vector y is called the projection of x in M and the operator P given by $Px = y$ is called the projection on M. We also denote this operator by P_M. Now, we shall see that P is a symmetric operator on H. In fact, let $x, y \in H$, write $x = x_1 + x_2$, $y = y_1 + y_2$ where $x_1, y_1 \in M$, $x_2, y_2 \in M^{\perp}$. We have also $(Px, y) = (x_1, y_1 + y_2) = (x_1, y_1)$ and $(x, Py) = (x_1 + x_2, y_1) = (x_1, y_1)$.

Other properties of the projection P: P is a linear operator, this follows from the uniqueness of the decompositions $x = x_1 + x_2$, where $x_1 \in M$ and $x_2 \in M^{\perp}$. Also, $P^2 = P$, due to the fact that an element in M is written as $x = x + \mathbf{0}$. Note also that operator $Q = I - P$ is simply the projection on the subspace $M^{\perp}$. A simple estimate is valid for (orthogonal) projections, namely: $\|P\|_{\mathcal{L}(H)} \le 1$; this is obtained as follows: $\forall x \in H$, $\|x\|^2 = \|Px\|^2 + \|Qx\|^2$, hence $\|Px\| \le \|x\|$. The order relations: $\Theta \le P \le I$ hold true, in the sense that, $\forall x \in H$, $(Px,x) = (P^2x, x) = (Px, Px) \ge 0$; as well as: $\forall x \in H$, $(Px, x) \le \|Px\|\,\|x\| \le \|x\|^2 = (Ix, x)$.

Note that if $M = H$, $P_M = I$(dentity); if $M = \{\mathbf{0}\}$, $P_M = \Theta$. If $P \ne \Theta$ there is a point $x \in M = P(H)$, where $x \ne 0$. Then $\|Px\| = \|x\|$ which gives that $\|P\| = 1$.

Now, there is a converse to these statements, which we present here as:

Theorem 1.1 *Let P be a linear continuous symmetric operator on the Hilbert space H, such that $P^2 = P$. Then P is the orthogonal projection on the space $P(H)$.*

Proof It is readily seen that $M = P(H) = \{Px,\ x \in H\}$ is a linear set in H. It is also a closed set being in fact the kernel of the continuous operator $I - P$: $y \in M \leftrightarrow y - Py = \mathbf{0}$. Next, note that: $(x - Px, Py) = \left(Px - P^2 x, y\right) = \mathbf{0}$, $\forall y \in H$, so that $x - Px \in M^{\perp}$. Therefore: $x = Px + (x - Px)$ is the unique decomposition of x as a sum $y + z$, with $y \in M$ and $z \in M^{\perp}$. This shows that P is the projection on M. ■

Next, let us prove the following:

In the Hilbert space H, consider two (closed) linear subspaces M and N with their associated projections P_M, P_N. Assume that $P_M P_N = P_N P_M$. Then $P_N P_M$ is the orthogonal projection on $M \cap N$.

We apply the previous theorem. Obviously, $P_N P_M$ is linear continuous in H. Also $(P_N P_M x, y) = (P_M x, P_N y) = (x, P_M P_N y) = (x, P_N P_M y)$, $\forall x, y \in H$. Thus $P_N P_M$ is also symmetric. Now, squaring $P_N P_M$ we obtain:

$$\left(P_N P_M\right)^2 = P_N P_M P_N P_M = P_N\left(P_N P_M\right) P_M = P_N^2 P_M^2 = P_N P_M .$$

Therefore $P_N P_M$ is the orthogonal projection on $\left(P_N P_M\right)(H)$. The last thing to establish is that $\left(P_N P_M\right)(H) = M \cap N$. Take first $x \in M \cap N$; then $P_M x = x$ and $P_N\left(P_M x\right) = P_N x = x$, so that $x \in \left(P_N P_M\right)(H)$. Next, let $y \in \left(P_N P_M\right)(H)$; so $y = P_N P_M x$ and $y \in P_N(H) = N$; also $y = P_M P_N x$ and $y \in P_M(H) = M$. Thus, $y \in M \cap N$.

Definition 1.1 *Two projections P_1, P_2 are orthogonal if $P_1 P_2 = \mathbf{0}$.*

Remark We have, $\forall x, y \in H$, $\left(P_2 P_1 x, y\right) = \left(P_1 x, P_2 y\right) = \left(x, P_1 P_2 y\right) = 0$. Thus, taking $y = P_2 P_1 x$, we find $P_2 P_1 x = \mathbf{0}$, $\forall x \in H$, hence $P_2 P_1 = \mathbf{0}$. Thus P_1, P_2 are orthogonal iff $P_1 P_2 = P_2 P_1 = \mathbf{0}$.

We give the following:

Theorem 1.2 *Two projections P_M, P_N are orthogonal iff $M \perp N$.*

In fact, if $M \perp N$ and $x, y \in H$, we get $0 = \left(P_M x, P_N y\right) = \left(x, P_M P_N y\right)$. Thus, $P_M P_N y = \mathbf{0}$, $\forall y \in H$, and P_M, P_N are orthogonal.

Conversely, let $P_M P_N = \mathbf{0}$ and $z \in M$, $u \in N$. Then $z = P_M z$, $u = P_N u$ and accordingly $(z,u) = (P_M z, P_N u) = 0$. ■

Next we have

Theorem 1.3 *Consider two orthogonal projections P_M and P_N. Then $P_M P_N = P_N P_M = P_M$ iff $M \subseteq N$.*

It is convenient to have the following:

Lemma *If A,B are linear continuous operators in the Hilbert space H, then $(AB)^* = B^* A^*$.*

In fact, $\forall x, y \in H$, we have $((AB)x, y) = (Bx, A^* y) = (x, B^* A^* y) = (x, (AB)^* y)$. This gives the lemma. ■

Proof of Theorem 1.3 Let $P_M P_N = P_M$; then $P_N P_M = P_M^* = P_M$ too (remember that orthogonal projections are symmetric and thus coincide with their adjoint). Take now $x \in M$; then $P_M x = x$ and $P_N(P_M x) = P_N x \in N$, hence $M \subseteq N$.

Conversely, let $M \subseteq N$. Then $P_N(P_M x) = P_M x$ as P_N is the identity operator on N. ■

At this stage it is convenient to introduce:

Definition 1.2 *A projection P_L is a part of a projection P_M if $L \subseteq M$.*

Theorem 1.4 *A projection P_M is a part of a projection P_N iff $\|P_M x\| \leq \|P_N x\|$, $\forall x \in H$.*

In fact, if we apply Theorem 1.3 we get $M \subseteq N$ iff $P_M P_N = P_N P_M = P_M$. Then $\|P_M x\| = \|P_M P_N x\| \leq \|P_M\| \|P_N x\| \leq \|P_N x\|$, $\forall x \in H$.

Conversely, assume $\|P_M x\| \leq \|P_N x\|$, $\forall x \in H$, and prove that $M \subset N$. If not, $\exists x_0 \in M$, $x_0 \notin N$. Let $x_0 = y_0 + z_0$, $y_0 \in N$, $z_0 \perp N$, $z_0 \neq \mathbf{0}$. Then: $\|P_M x_0\|^2 = \|x_0\|^2 = \|y_0\|^2 + \|z_0\|^2 > \|y_0\|^2 = \|P_N x_0\|^2$, a contradiction. ■

Definition 1.3 *Let M,N be linear closed subspaces of the Hilbert space H, and $M \subset N$. Then, the 'orthogonal difference': $N \ominus M$ is the set: $\{x \in N, x \perp M\}$.*

Now we give:

Theorem 1.5 *With the assumptions in Definition 1.3, the equality*

$$N \ominus M = (P_N - P_M)(H)$$

holds true.

Proof Note first that $P_N - P_M$ is a linear continuous symmetric operator on H. Furthermore, squaring we get: $(P_N - P_M)(P_N - P_M) = P_N - P_M P_N - P_N P_M + P_M =$ (using Theorem 1.3) $P_N - P_M - P_M + P_M = P_N - P_M$. Thus, $P_N - P_M$ is also a projector. Next, take $x \in (P_N - P_M)(H)$, so $x = P_N z - P_M z \in N$. Take any element of M: $h = P_M u$. We get: $(P_N z - P_M z, P_M u) = (P_M P_N z - P_M z, u) = (P_M z - P_M z, u) = 0$. It follows that $(P_N - P_M)(H) \subset N \ominus M$.

Conversely, let $x \in N$, $x \perp M$. Then $x = P_N z$ and then $P_M z = P_M P_N z = P_M x = \mathbf{0}$ (as $x \perp M$). Hence $x = P_N z - P_M z \in (P_N - P_M)(H)$. ■

Definition 1.4 *Let A,B be continuous symmetric operators on the Hilbert space H. We say* $A \geq B$ *iff* $A - B \geq \Theta$ *iff* $(Ax, x) \geq (Bx, x)$, $\forall x \in H$.

Theorem 1.6 *Let* P_M, P_N *be projections on the subspaces M,N of the Hilbert space H. Then* $M \subset N$ *iff* $P_N \geq P_M$.

Proof Assume $M \subset N$; as seen in Theorem 1.5, $P_N - P_M$ is a projection, hence $P_N - P_M \geq \Theta$. Conversely, note first that if $P_N \geq P_M$, then $I - P_N \leq I - P_M$ and therefore $(\forall x \in H)$

$$\begin{aligned}\|(I - P_N)P_M x\|^2 &= ((I - P_N)P_M x,\ (I - P_N)P_M x) \\ &= ((I - P_N)^2 P_M x,\ P_M x) \\ &= ((I - P_N)P_M x,\ P_M x) \leq ((I - P_M)P_M x,\ P_M x) = 0\ .\end{aligned}$$

Thus $(I - P_N)P_M x = \mathbf{0} = P_M x - P_N P_M x$, and $P_N P_M = P_M$. Finally, apply Theorem 1.3. ■

2. In this section we discuss the existence and uniqueness of the positive symmetric square root of a given, positive symmetric operator. Thus, we consider in the Hilbert space H an operator A, $H \to H$, which is symmetric and positive: $(Ax, x) \geq 0$, $\forall x \in H$. We consider the equation $X^2 = A$, and look for a symmetric positive solution X.

Let us first note:

Proposition 2.1 *It is sufficient to assume* $\Theta \leq A \leq I$.

In fact, as A is continuous in H, we have $(Ax, x) \leq \|A\| \, \|x\|^2 = (\|A\|x; x)$, $\forall x \in H$ which means that $A \leq \|A\| I$, and $\frac{1}{\|A\|} A \leq I$. If X is a solution - symmetric and positive - to the equation $X^2 = \frac{1}{\|A\|} A$, then $Y = \|A\|^{\frac{1}{2}} X$ is a solution to $Y^2 = A$ which is again symmetric and positive. ■

Proposition 2.2 *Let* $B = I - A$ *and* Y *be a solution to the equation* $Y = \frac{1}{2}\left(B + Y^2\right)$. *Then* $X = I - Y$ *is a solution to* $X^2 = A$.

In fact $X^2 = (I - Y)^2 = I - 2Y + Y^2 = I - \left(B + Y^2\right) + Y^2 = I - B = A$.

We intend to solve the equation $Y = \frac{1}{2}\left(B + Y^2\right)$ by 'successive approximations'. We start with the operator $Y_0 = \Theta$ and then, inductively, define operators Y_n by the recurrence relations

$$Y_{n+1} = \frac{1}{2}\left(B + Y_n^2\right). \tag{2.1}$$

We note first that each Y_n is a polynomial in the operator B with positive coefficients:

$$Y_n = \sum_{p=1}^{p_n} a_p B^p, \quad \text{where } a_p \geq 0 \text{ (for } n = 1, 2, \ldots). \tag{2.2}$$

In fact, if Y_n has this form then $Y_{n+1} = \frac{1}{2}\left(B + \left(\sum_{p=1}^{p_n} a_p B^p\right)^2\right)$ which is a polynomial of the same kind.

Next, we shall see that, $\forall n = 1, 2, \ldots$, the operators $Y_{n+1} - Y_n$ are of the same form (2.2). Note in fact that $Y_{n+1} - Y_n = \frac{1}{2}\left(B + Y_n^2\right) - \frac{1}{2}\left(B + Y_{n-1}^2\right) = \frac{1}{2}\left(Y_n^2 - Y_{n-1}^2\right)$.

This last difference equals $\frac{1}{2}(Y_n - Y_{n-1})(Y_n + Y_{n-1})$ in view of the commutativity of all operators Y_n (from (2.2)). Thus we obtained

$$Y_{n+1} - Y_n = \frac{1}{2}(Y_n + Y_{n-1})(Y_n - Y_{n-1}) . \tag{2.3}$$

If, by induction, one assumes that $Y_n - Y_{n-1}$ is also of the form (2.2), then (2.3) will imply - together with (2.2) - that $Y_{n+1} - Y_n$ is of the same form.

Next, note that any polynomial $\sum_{1}^{k} a_p B^p$, where $a_p \geq 0$ and $B \geq \Theta$, is also a positive operator, and that the operator $B = I - A$ is positive. Then $\left(B^{2m+1}x, x\right) = \left(B \cdot B^m x, B^m x\right) \geq 0$, $\forall x \in H$ and $B^{2m+1} \geq \Theta$, $\forall m \in \mathbb{N}$. (For even powers we get in any case $\left(B^{2m}x, x\right) = \left(B^m x, B^m x\right) \geq 0$.)

Therefore we obtain:

$$Y_n \geq 0 \quad \forall n \in \mathbb{N}, \quad \text{and} \quad Y_n \geq Y_{n-1}, \quad \forall n \in \mathbb{N} . \tag{2.4}$$

Still by induction we see that

$$\|Y_n\| \leq 1 , \quad \forall n \in \mathbb{N} . \tag{2.5}$$

In fact, $Y_1 = \frac{1}{2}B$ gives $\|Y_1\| = \frac{1}{2}\|B\|$.

Actually, $\|B\| \leq 1$ (this follows from a general, well-known result - see for instance Young [14], p. 79):

If T is a Hermitian operator on a Hilbert space H then $\|T\| = \sup_{\|x\|=1} |(Tx, x)|$.

Therefore $\|Y_1\| \leq \frac{1}{2}$.

Assume now again - by induction - that $\|Y_n\| \leq 1$; then $Y_{n+1} = \frac{1}{2}\left(B + Y_n^2\right)$ and

$$\|Y_{n+1}\| \leq \frac{1}{2}\left(\|B\| + \|Y_n\|^2\right) \leq \frac{1}{2}(1+1) = 1 . \tag{2.6}$$

From (2.4)-(2.5) we infer that the sequence of (symmetric) operators $(Y_n)_1^\infty$ is increasing:

$$\Theta \le Y_1 \le Y_2 \le \ldots \le I \ , \quad \text{and} \quad \|Y_n\| \le 1 \quad \forall n = 1,2,\ldots \ . \tag{2.7}$$

We now state (and prove) a more general result:

Lemma 2.1 *Consider in the Hilbert space H the sequence of continuous symmetric operators $(A_n)_1^\infty$, where*

$$A_1 \le A_2 \le \ldots \le T \tag{2.8}$$

with (another) continuous symmetric operator T. Then, there exists a linear continuous operator A, $H \to H$, A symmetric, and

$$\lim_{n\to\infty} A_n x = Ax \ , \quad \forall x \in H \ . \tag{2.9}$$

Proof It is obvious that the numerical sequence $((A_n x, x))_1^\infty$ - where $x \in H$ - is increasing and bounded from above. It is accordingly a Cauchy sequence.

Note now the following extension of the Cauchy-Schwarz inequality (1.10)-I

If $A \in \mathcal{L}(H)$ is symmetric and $\ge \Theta$, then

$$|(Ax,y)| \le (Ax,x)^{\frac{1}{2}} (Ay,y)^{\frac{1}{2}} \tag{2.10}$$

(we get (1.10)-I for A = Identity).

Take in fact, $\forall \lambda \in \mathbb{R}$, $z_\lambda = x + \lambda (Ax,y) y$. Then $(Az_\lambda, z_\lambda) \ge 0$ as also

$$\begin{aligned}
(Az_\lambda, z_\lambda) &= (A(x+\lambda(Ax,y)y),\ x+\lambda(Ax,y)y) \\
&= (Ax + \lambda(Ax,y)Ay,\ x+\lambda(Ax,y)y) \\
&= (Ax,x) + \lambda(Ax,y)(Ay,x) + (Ax, \lambda(Ax,y)y) + (\lambda(Ax,y)Ay, \lambda(Ax,y)y) \\
&= (Ax,x) + \lambda(Ax,y)(y,Ax) + \lambda\overline{(Ax,y)}(Ax,y) + \lambda^2 |(Ax,y)|^2 (Ay,y) \\
&= (Ax,x) + 2\lambda |(Ax,y)|^2 + \lambda^2 |(Ax,y)|^2 (Ay,y) \\
&\ge 0 \ , \quad \forall \lambda \in \mathbb{R} \ .
\end{aligned}$$

This implies, as usual, that $|(Ax,y)|^4 - |(Ax,y)|^2 (Ax,x)(Ay,y) \le 0$ which is (2.10).

We now continue the **Proof** of **Lemma 2.1**. We have, using the notation $A_{n,m} = A_n - A_m$, that - because of (2.10) - (for $n \ge m$),

$$\|A_{n,m}x\|^2 = (A_{n,m}x, A_{n,m}x) \le (A_{n,m}x, x)^{\frac{1}{2}} (A_{n,m}A_{n,m}x, A_{n,m}x)^{\frac{1}{2}}$$

$$\le (A_{n,m}x, x)^{\frac{1}{2}} \|A_{n,m}\|^{\frac{1}{2}} \|A_{n,m}x\|$$

and

$$\|A_{n,m}x\| \le (A_{n,m}x, x)^{\frac{1}{2}} \|A_{n,m}\|^{\frac{1}{2}}. \tag{2.11}$$

On the other hand for $n > m$,

$$\|A_{n,m}\| = \sup_{\|x\|\le 1} |((A_n - A_m)x, x)|$$

$$\le \sup_{\|x\|\le 1} |(A_n x, x)| + \sup_{\|x\|\le 1} |(A_m x, x)| \le C = 2\max(\|A_1\|, \|T\|)$$

(as follows from $(A_1 x, x) \le (A_n x, x) \le (Tx, x)\ \forall n \in \mathbb{N},\ \forall x \in H$). Therefore, (2.11) implies the estimate

$$\|A_{n,m}x\| \le C(A_{n,m}x, x)^{\frac{1}{2}} \tag{2.12}$$

and as the sequence $((A_n x, x))$ is a Cauchy sequence, the same is true for the sequence $(A_n x)_1^\infty$ in H.

Therefore, $\lim_{n \to \infty} A_n x$, denoted by Ax, will exist $\forall x \in H$. It is also a symmetric operator: $\forall x, y \in H$, $(Ax, y) = \lim_{n\to\infty} (A_n x, y) = \lim_{n\to\infty} (x, A_n y) = (x, Ay)$, hence – as previously seen – $A \in \mathcal{L}(H)$.

Lemma 2.1 is, accordingly, completely proved. ■

We now pursue the **proof of Proposition 2.2**.

From (2.7) and Lemma 2.1 we infer the existence of a symmetric (continuous operator) Y, $H \to H$, such that

$$\lim_{n\to\infty} Y_n x = Yx \,, \qquad \forall x \in H \,. \tag{2.13}$$

We know that

$$Y_{n+1}x = \frac{1}{2}\left(Bx + Y_n^2 x\right), \qquad \forall x \in H \,. \tag{2.14}$$

We see also that $YY_m = Y_m Y$, $\forall m \in \mathbb{N}$. In fact, $\forall x \in H$, we obtain:

$$YY_m x = \lim_{n\to\infty} Y_n\left(Y_m x\right) = \lim_{n\to\infty} Y_m\left(Y_n x\right) = Y_m Yx \tag{2.15}$$

(the operators Y_m commute between them, as polynomials of the same symmetric operator B).

Then, we have $Y_n^2 - Y^2 = (Y_n - Y)(Y_n + Y)$ and

$$\begin{aligned}\left\|Y_n^2 x - Y^2 x\right\| &= \left\|(Y_n + Y)(Y_n x - Yx)\right\| \\ &\le \left(\|Y_n\| + \|Y\|\right)\|Y_n x - Yx\| \le \left(1 + \|Y\|\right)\|Y_n x - Yx\| \,, \qquad \forall x \in H\end{aligned}$$

which shows that

$$\lim_{n\to\infty} Y_n^2 x = Y^2 x \,, \qquad \forall x \in H \,. \tag{2.16}$$

Thus, from (2.14) we now derive that:

$$\lim_{n\to\infty} Y_{n+1}x = Yx = \frac{1}{2}\left(Bx + Y^2 x\right). \tag{2.17}$$

Therefore $Y = \frac{1}{2}\left(B + Y^2\right)$ and then (Proposition 2.2), $X = I - Y$ is a solution to $X^2 = A$.

■

We can also establish the uniqueness of the positive symmetric square root. (The proof uses the 'existence' part which is somewhat unusual.)

Proposition 2.3 *In the Hilbert space H, let A, $H \to H$, be symmetric and positive, and C any positive symmetric square root of A. Then $C = X$, the square root found in Proposition 2.1-2.2.*

In fact, let us note first that $AC = CA$ and $CX = XC$: We have $A = C^2$, hence $AC = C^3$, $CA = C^3$. Next, we have $Xx = \lim_{n\to\infty} Y_n x$, $\forall x \in H$, where Y_n is - for any n - a polynomial in $B = I - A$, hence a polynomial in A, hence it commutes with A. Then, $\forall x \in H$, $CXx = C(\lim Y_n x) = \lim C(Y_n x)$. Now $CY_n = Cp(A) = p(A)C$. Also $XCx = \lim Y_n Cx$, where $Y_n C = p(A)C = CY_n$. Thus $XCx = \lim CY_n x = CXx$, $\forall x \in H$. We found that C commutes with A and with X. We shall now see that $Cx = Xx$, $\forall x \in H$. Denote $y = (X - C)x$. We see that $X^2 = C^2 = A$, hence

$$0 = \left((X^2 - C^2)x, y\right) = ((X + C)(X - C)x, y) = ((X + C)y, y) = (Xy, y) + (Cy, y) \ . \quad (2.18)$$

On the other hand, both operators X and C possess positive symmetric square roots, which we denote by X' and C'. We get then

$$(Xy, y) = (X'X'y, y) = \|X'y\|^2 \quad \text{and} \quad (Cy, y) = \|C'y\|^2 \ . \quad (2.19)$$

From (2.18) we then infer that $0 = \|X'y\|^2 + \|C'y\|^2$, so that $X'y = C'y = \mathbf{0}$ and therefore $X'(X'y) = C'(C'y) = \mathbf{0}$ which means that $Xy = Cy = \mathbf{0}$.

The final step is:

$$\begin{aligned} \|y\|^2 &= \|(X - C)x\|^2 = ((X - C)x,\ (X - C)x) = \left((X - C)^2 x, x\right) \\ &= ((X - C)y, x) = (Xy, x) - (Cy, x) = 0 \ . \end{aligned} \quad (2.20)$$

This shows that $C = X$. ■

Thus, from now on, we are entitled to denote the unique positive symmetric square root of the symmetric positive operator A by $A^{1/2}$.

We note (to end this chapter) the consequences of the above results.

Corollary 2.1 *Let A_1, A_2 be positive symmetric operators in H, such that $A_1 A_2 = A_2 A_1$. Then $A_1 A_2$ is again a symmetric positive operator in H.*

The symmetry of $A_1 A_2$ is readily derived – without the help of the square root

$$(A_1 A_2 x, y) = (A_2 x, A_1 y) = (x, A_2 A_1 y) = (x, A_1 A_2 y) \ , \qquad \forall x, y \in H \ .$$

Next, note that, as A_1 commutes with A_2 it also commutes with $A_2^{1/2}$. Then $(A_1A_2x,x)=\left(A_1A_2^{1/2}A_2^{1/2}x,x\right)=\left(A_2^{1/2}A_1A_2^{1/2}x,x\right)=\left(A_1A_2^{1/2}x,A_2^{1/2}x\right)\geq 0$, as A_1 is positive.

Corollary 2.2 *Let* $A_1 \geq A_2$ *be two symmetric operators in H and let* $A_3 \geq 0$ *be another symmetric operator which commutes with* A_1 *and with* A_2. *Then* $A_1A_3 \geq A_2A_3$.

In fact, $A_1A_3 - A_2A_3 = (A_1 - A_2)\cdot A_3$. Here $A_1 - A_2 \geq \Theta$, $A_3 \geq \Theta$ and $A_1 - A_2$ commutes with A_3. We finally apply Corollary 2.1.

Chapter VIII

Semidynamical systems and C_0-semigroups

Introduction

This chapter is a continuation of Chapter V. We complete the results given there with some new definitions and properties.

1. Here we assume an elementary knowledge of the concept of a 'metric space' (see for instance [22]). We shall follow the presentation of ([12]-Chapter III).

Definition 1.1 A semidynamical system on a metric space $\mathcal{X}$ is a mapping u: $[0,\infty)\times\mathcal{X}\to\mathcal{X}$ enjoying the following properties:

(i) $\forall x\in\mathcal{X}$, the function $t\to u(t,x)$ is continuous from $[0,\infty)$ into $\mathcal{X}$.

(ii) $\forall t\geq 0$, the function $x\in\mathcal{X}\to u(t,x)\in\mathcal{X}$ is continuous.

(iii) $u(0,x)=x$, $\forall x\in\mathcal{X}$.

(iv) For all $t,s\in[0,\infty)$ and $x\in\mathcal{X}$, the equality

$$u(t+s,x)=u(t,u(s,x))$$

holds true.

In connection with this definition, the *motion* originating at $x\in\mathcal{X}$ is defined as the mapping $t\to u(t,x)$, $[0,\infty)\to\mathcal{X}$.

A point $x_e\in\mathcal{X}$ is called an equilibrium point if $u(t,x_e)=x_e$, $\forall t\geq 0$ (it is also called a 'rest-point').

A set $S\subset\mathcal{X}$ is invariant under u if $x\in S\Rightarrow u(t,x)\in S$, $\forall t\geq 0$. Occasionally, we refer to $\mathcal{X}$ as the *state space* and then $u(t,x)$ is the state at 'time' $t\geq 0$.

A closely related concept is now given as:

Definition 1.2 For a metric space $\mathcal{X}$, a family $\{S(t)\}_{t\geq 0}$ of continuous functions, $\mathcal{X} \to \mathcal{X}$, is a strongly continuous semigroup of functions (also called 'operators') if

(i) the function $t \to S(t)x \in \mathcal{X}$ is continuous $\forall t \geq 0$;

(ii) $S(0)x = x, \ \forall x \in \mathcal{X}$;

(iii) $S(t+\sigma) = S(t)S(\sigma), \ \forall t, \sigma \in [0, \infty)$.

As we see immediately, every family $\{S(t)\}_{t\geq 0}$ satisfying (i)-(iii) determines a semidynamical system when one puts

$$u(t,x) = S(t)x\,, \qquad t \geq 0,\ x \in \mathcal{X} \tag{1.1}$$

(for instance, $u(t+\sigma, x) = S(t+\sigma)x = S(t)S(\sigma)x = u(t, S(\sigma)x) = u(t, u(\sigma, x))$).

Conversely, if $u(t,x)$ is a semidynamical system, the operators $S(t)$, $\mathcal{X} \to \mathcal{X}$ defined again by (1.1) form - obviously - a strongly continuous semigroup on the metric space $\mathcal{X}$.

We now take a very special case: $\mathcal{X}$ is a Banach space with the usual metric, $d(x,y) = \|x-y\|$ and the continuous functions $S(t)$ in Definition 1.2 are also linear operators (thus $S(t) \in \mathcal{L}(\mathcal{X})$, $\forall t \geq 0$). We retrieve the semigroups of class C_0 in Chapter V.

The following result could be of interest.

Proposition 1.1 *In the Banach space $\mathcal{X}$, let A, $\mathcal{D}(A) \subset \mathcal{X} \to \mathcal{X}$ be the infinitesimal generator of the C_0-semigroup $S(t)$. Then, $\forall \alpha \in \mathbb{R}$, the family $\left\{e^{\alpha t} S(t)\right\}_{t\geq 0}$ is again a C_0-semigroup on $\mathcal{X}$ with infinitesimal generator $\alpha I + A$.*

Proof First, $\forall x \in \mathcal{X}$, the function $t \geq 0 \to e^{\alpha t} S(t)x$ is the product of two continuous functions: $t \to e^{\alpha t}$, $\mathbb{R}^+ \to \mathbb{R}$, and $t \to S(t)x$, $\mathbb{R}^+ \to \mathcal{X}$. It is accordingly, a continuous function, $\mathbb{R}^+ \to \mathcal{X}$.

Next, for $t = 0$, $e^{\alpha t} S(t)x = x$. Finally:

$$e^{\alpha(t+s)} S(t+s) = e^{\alpha t} e^{\alpha s} S(t)S(s) = e^{\alpha t} S(t) \cdot e^{\alpha s} S(s)\,.$$

Thus, properties (a)-(c) in Chapter V-1 are established.

Let us denote by $T(t)$ the semigroup $e^{\alpha t} S(t)$. We shall compute its infinitesimal generator: we have, $\forall \eta > 0$, $x \in \mathcal{X}$,

$$\frac{T(\eta)x-x}{\eta} = \frac{e^{\alpha\eta}S(\eta)x-x}{\eta} = e^{\alpha\eta}\frac{S(\eta)x-x}{\eta} + \frac{e^{\alpha\eta}-1}{\eta}x \tag{1.2}$$

hence, for $x \in \mathcal{D}(A)$

$$\lim_{\substack{\eta\to 0\\ \eta>0}} \frac{T(\eta)x-x}{\eta} = Ax+\alpha x = (\alpha I + A)x \ . \tag{1.3}$$

On the other hand, if the limit in the left-hand side of (1.2) exists, then necessarily $\lim\limits_{\eta\to 0}\dfrac{S(\eta)x-x}{\eta}$ will exist too. ∎

In many instances and applications it is of fundamental importance to know if a linear operator happens to be an infinitesimal generator. Obviously, by the results in Chapter V, this operator must be closed and densely defined on $\mathcal{X}$.

We now prove:

Proposition 1.2 *If A is linear continuous operator on the Banach space $\mathcal{X}$, then A is an infinitesimal generator.*

The proof is done by an explicit construction of a (C_0) semigroup which has A as infinitesimal generator. We consider the series of operators

$$\sum_0^\infty \frac{t^n}{n!}A^n = \lim_{N\to\infty}\left(\sum_0^N \frac{t^n}{n!}A^n\right) = I + \sum_1^\infty \frac{t^n}{n!}A^n \ . \tag{1.4}$$

Obviously, each power A^n is an operator in $\mathcal{L}(X)$. The finite sums $B_N = \sum\limits_0^N \dfrac{t^n}{n!}A^n$ are also in $\mathcal{L}(X)$, $\forall N = 1,2,\ldots$, $\forall t \geq 0$. Furthermore, the sequence (in $\mathcal{L}(X)$) $(B_N)_1^\infty$, is a Cauchy sequence: this is a consequence of the estimate, for $N > M$,

$$\left\|B_N - B_M\right\|_{\mathcal{L}(X)} = \left\|\sum_{M+1}^N \frac{t^n}{n!}A^n\right\|_{\mathcal{L}(X)} \leq \sum_{M+1}^N \frac{t^n}{n!}\|A\|^n \tag{1.5}$$

and of the well-known fact that the numerical series $\sum\limits_0^\infty \dfrac{\lambda^n}{n!}$ is convergent $\forall \lambda \in \mathbb{R}$.

Therefore, the limit of $(B_N)_1^\infty$ in $\mathcal{L}(X)$ will exist ($\forall t \geq 0$); we denote it by $e^{At} = \exp(At)$. Note also the estimate

$$\|B_N\| \leq \sum_0^N \frac{t^n}{n!}\|A\|^n \leq e^{\|A\|t}, \qquad t \geq 0,\ N = 1,2,\dots . \tag{1.6}$$

As $B_N \to e^{At}$, it follows that $\|B_N\|_{\mathcal{L}(X)} \to \|e^{At}\|_{\mathcal{L}(X)}$. Then, from (1.6) we find

$$\|e^{At}\|_{\mathcal{L}(X)} \leq \exp(t\|A\|), \qquad \forall t \geq 0 . \tag{1.7}$$

Concerning the continuity of the function $t \geq 0 \to \exp(tA)x$, $\forall x \in \mathcal{X}$, we see that more is true, namely the continuity of the function: $t \to \exp(tA)$, from $[0,\infty)$ in $\mathcal{L}(X)$.

First of all we prove the continuity for $t = 0$: as $\exp(tA) = I$(dentity) for $t = 0$ we consider the estimate

$$\|\exp(tA) - I\| = \left\|\sum_1^\infty \frac{t^n}{n!}A^n\right\| \leq \sum_1^\infty \frac{t^n}{n!}\|A\|^n = e^{t\|A\|} - 1, \qquad t \geq 0 \tag{1.8}$$

whence $\lim\limits_{\substack{t\to 0\\ t>0}}\|\exp(tA) - I\| = 0$.

Before proceeding with the proof of continuity for all $t > 0$, we shall establish the 'semigroup property' for the family $S(t) = \exp(tA)$, namely the relation

$$S(t+\sigma) = S(t)S(\sigma), \qquad \forall t,\sigma \in [0,\infty) . \tag{1.9}$$

We give a preliminary result interesting in itself:

Lemma 1.1 *Consider two series of elements in* $\mathcal{L}(X)$: $\sum\limits_{n=0}^\infty A_n$ *and* $\sum\limits_{n=0}^\infty B_n$. *We assume 'absolute convergence' of the first series, that is:* $\sum\limits_0^\infty \|A_n\| < \infty$ *and convergence of the second series (thus,* $\exists B \in \mathcal{L}(X)$*, and* $\lim\limits_{N\to\infty}\sum\limits_0^N B_n = B$ *in* $\mathcal{L}(X)$*).*

Define elements C_n *in* $\mathcal{L}(X)$ *by the formula*

$$C_n = A_0 B_n + A_1 B_{n-1} + \ldots + A_n B_0 = \sum_{k=0}^{n} A_k B_{n-k} \ . \tag{1.10}$$

Then $\sum_0^\infty C_n = A \cdot B$ *where* $A = \sum_0^\infty A_n$.

Proof of Lemma Note first that, $\forall p \in \mathbb{N}$, $\sum_{k=0}^{p} C_k = \sum_{k=0}^{p} A_k \left(\sum_{j=0}^{p-k} B_j \right)$. Therefore we obtain

$$\sum_{k=0}^{p} C_k = \sum_{k=0}^{p} A_k \left(\sum_{j=0}^{\infty} B_j - \sum_{j=p-k+1}^{\infty} B_j \right) = \left(\sum_{k=0}^{p} A_k \right) B - \sum_{k=0}^{p} A_k \left(\sum_{j=p-k+1}^{\infty} B_g \right).$$

Define $R_m = \sum_{j=m+1}^{\infty} B_j$; then $\sum_{j=p-k+1}^{\infty} B_j = R_{p-k}$ and

$$\sum_{k=0}^{p} C_k = \left(\sum_{k=0}^{p} A_k \right) B - \sum_{k=0}^{p} A_k R_{p-k} \ . \tag{1.11}$$

The main step now: we obtain

$$\lim_{p \to \infty} \sum_{k=0}^{p} A_k R_{p-k} = \Theta \ . \tag{1.12}$$

In fact, $R_m \to \Theta$ as $m \to \infty$, hence, $\forall \varepsilon > 0$, $\exists m_\varepsilon \in \mathbb{N}$ such that for $m > m_\varepsilon$ we get

$$\|R_m\| < \frac{\varepsilon}{2A^*} , \quad \text{with } A^* = \sum_0^\infty \|A_n\| \ . \tag{1.13}$$

Let us define now the number $L = \max\left(\|R_0\|, \|R_1\|, \ldots, \|R_{m_\varepsilon}\|, \frac{\varepsilon}{2A^*} \right)$. Then (as seen from (1.1)) one obtains that $\|R_m\| \le L$, $\forall m \in \mathbb{N}$.

On the other hand, there exists $n_\varepsilon > m_\varepsilon$ such that, for $n_\varepsilon < n \le p$,

$$\sum_{k=n}^{p} \|A_k\| < \frac{\varepsilon}{2L} \ . \tag{1.14}$$

Let us now take $p \geq 2n_\varepsilon$: then, for $0 \leq k \leq n_\varepsilon$ we have $p-k \geq n_\varepsilon > m_\varepsilon$ and consequently one gets estimates:

$$\left\| \sum_{k=0}^{p} A_k R_{p-k} \right\| \leq \left\| \sum_{k=0}^{n_\varepsilon} A_k R_{p-k} \right\| + \left\| \sum_{k=n_\varepsilon+1}^{p} A_k R_{p-k} \right\|$$

$$\leq \frac{\varepsilon}{2A^*} \sum_{k=0}^{n_\varepsilon} \|A_k\| + L \sum_{n_\varepsilon+1}^{p} \|A_k\| \leq \frac{\varepsilon}{2A^*} A^* + L \cdot \frac{\varepsilon}{2L} = \varepsilon \ . \tag{1.15}$$

Therefore (1.12) holds true and then from (1.11) we derive $\sum_{k=0}^{\infty} C_k$ exists and equals $\left(\sum_{k=0}^{\infty} A_k \right) \cdot \left(\sum_{k=0}^{\infty} B_k \right)$. ■

Now, using the lemma we obtain equality (1.9) in the form of the following

Corollary 1.1 *In the Banach space $\mathcal{X}$, let $T \in \mathcal{L}(X)$ and $t, s \in [0,\infty)$. Define operators $A_n = \frac{t^n}{n!} T^n$, $B_n = \frac{s^n}{n!} T^n$. Then $\sum_{n=0}^{\infty} A_n = \exp(tT)$, $\sum_{n=0}^{\infty} B_n = \exp(sT)$ with $\sum_{0}^{\infty} \|A_n\| < \infty$, $\sum_{0}^{\infty} \|B_n\| < \infty$.*

Therefore $\left(\sum_{0}^{\infty} A_n \right) \cdot \left(\sum_{0}^{\infty} B_n \right) = \sum_{0}^{\infty} C_n$, where C_n is given in (1.10).

We now compute C_n explicitly: we have

$$C_n = I \frac{s^n T^n}{n!} + \frac{t}{1!} T \frac{s^{n-1} T^{n-1}}{(n-1)!} + \frac{t^2 T^2}{2!} \frac{s^{n-2} T^{n-2}}{(n-2)!} + \ldots + \frac{t^n}{n!} T^n I$$

$$= \frac{T^n}{n!} \left(s^n + n s^{n-1} t + \frac{n(n-1)}{2!} s^{n-2} t^2 + \ldots + t^n \right) = (s+t)^n \frac{T^n}{n!} \ .$$

Accordingly we get

$$\exp((s+t)T) = \exp(tT) \cdot \exp(sT) \ , \qquad \forall s,t \geq 0 \ . \tag{1.16}$$

■

Remark Compare with ([6], p. 84).

We are now ready to establish the continuity (in the $\mathcal{L}(X)$−sense) of the semigroup $S(t) = \exp(tA)$, for all $t > 0$. We have in fact (for small h)

$$S(t+h) - S(t) = \begin{cases} S(t)(S(h)-I) , & \text{if } \ h > 0 \\ -S(t+h)(S(-h)-I) , & \text{if } \ h < 0 \text{ and } t+h > 0 \end{cases}$$

hence

$$\|S(t+h - S(t)\| \le \begin{cases} \|S(t)\| \, \|S(h)-I\| , & \text{for } \ h > 0 \\ \|S(t+h)\| \, \|S(-h)-I\| , & \text{for } \ h < 0 . \end{cases} \tag{1.17}$$

We have already seen that:

$$\lim_{\substack{h\to 0 \\ h>0}} \|S(h) - I\| = 0 .$$

Furthermore, from (1.7) we derive that

$$\|S(t+h)\| \le \exp(t\|A\|)\exp(h\|A\|) \le \exp(t\|A\|) , \qquad \text{for } \ h < 0 .$$

Hence, in any case

$$\lim_{h\to 0} \|S(t+h) - S(t)\| = 0 .$$

It is therefore established that $\{S(t)\}_{t\ge 0}$ is a C_0-semigroup in $\mathcal{X}$. It remains to prove that the operator A is its infinitesimal generator. We have in fact (for $t \ge 0$)

$$\frac{1}{t}[S(t) - I] = \frac{1}{t}\sum_{1}^{\infty} \frac{t^n}{n!} A^n = A \sum_{m=0}^{\infty} \frac{t^m A^m}{(m+1)!}$$

hence

$$\frac{1}{t}[S(t) - I] - A = A\left[\sum_{m=0}^{\infty} \frac{t^m A^m}{(m+1)!} - I\right] = A\sum_{m=1}^{\infty} \frac{t^m A^m}{(m+1)!} = \sum_{m=1}^{\infty} \frac{t^m}{(m+1)!} A^{m+1}$$

$$= \sum_{n=2}^{\infty} \frac{t^{n-1}}{n!} A^n ,$$

so that

$$\left\|\frac{1}{t}[S(t)-I]-A\right\| \le \sum_{2}^{\infty} \frac{t^{n-1}}{n!}\|A\|^n = t\|A\|^2 \sum_{n=2}^{\infty} \frac{t^{n-2}}{n!}\|A\|^{n-2}$$

$$\le t\|A\|^2 \sum_{n=2}^{\infty} \frac{t^{n-2}}{(n-2)!}\|A\|^{n-2} = t\|A\|^2 \exp(t\|A\|) \to 0 \quad \text{as} \quad t \to 0 . \quad \blacksquare$$

Chapter IX

Uniqueness of weak solutions for a second-order abstract Cauchy problem in Hilbert space

Introduction

In this section we study - following the article by H.A. Levine [7] - the question of the uniqueness of a certain class of weak solutions to the equation: $u''(t) = Mu(t)$, where M is a symmetric operator and u takes values in a Hilbert space.

1. Let H be a Hilbert space over $\mathbb{R}$ (a 'real' Hilbert space) and $D \subseteq H$ be a dense linear subspace – usually non-closed. We let $(\; , \;)$ and $\| \; \|$ denote the scalar product and norm, respectively, on H. Consider (to start with) the second-order equation in H:

$$u''(t) = Mu(t) \, , \quad \text{on} \quad [0, \infty) \tag{1.1}$$

with Cauchy data:

$$u(0) = u_0 \, , \quad u'(0) = v_c \, . \tag{1.2}$$

M is a symmetric operator with domain D: thus

$$(Mx, y) = (x, My) \, , \quad \forall x, y \in D \, . \tag{1.3}$$

We associate with M a symmetric bilinear form $\mathcal{M}(x, y)$ on $D \times D$, via the formula

$$\mathcal{M}(x, y) = (Mx, y) \, , \quad \forall x, y \in D \, . \tag{1.4}$$

We see immediately from this definition that $\mathcal{M}$ is symmetric:

$$\mathcal{M}(x, y) = \mathcal{M}(y, x) = (My, x) \, , \quad \forall x, y \in D \tag{1.5}$$

and bilinear:

$$\mathcal{M}(\alpha x_1 + \beta x_2, y) = \alpha\mathcal{M}(x_1, y) + \beta\mathcal{M}(x_2, y)$$

as well as

$$\mathcal{M}(x, \alpha y_1 + \beta y_2) = \alpha\mathcal{M}(x, y_1) + \beta\mathcal{M}(x, y_2) .$$

We shall first prove that the solutions of (1.1)-(1.2) satisfy a certain relation involving a special class of 'test-functions'.

Precisely, we consider functions $\varphi(\cdot)$, $[0,\infty) \to D$, such that $\varphi(\cdot) \in C^1([0,\infty); H)$. Then, from (1.1) we first get:

$$(\varphi(t), u''(t)) = (\varphi(t), Mu(t)) = \mathcal{M}(\varphi(\cdot), u(\cdot)) . \tag{1.6}$$

In view of the simple formula

$$(\varphi(t), u''(t)) = \frac{\mathrm{d}}{\mathrm{d}t}(\varphi(t), u'(t)) - (\varphi'(t), u'(t)) \tag{1.7}$$

we get (from (1.6)) the relation

$$\frac{\mathrm{d}}{\mathrm{d}t}(\varphi(t), u'(t)) - (\varphi'(t), u'(t)) = \mathcal{M}(\varphi(\cdot), u(\cdot)) \tag{1.8}$$

whence, through integration from 0 to $t > 0$, one obtains

$$(\varphi(t), u'(t)) - (\varphi(0), v_0) = \int_0^t [(\varphi'(\eta), u'(\eta) + \mathcal{M}(\varphi(\eta), u(\eta))] \,\mathrm{d}\eta \tag{1.9}$$

which, in this Chapter, is the weak form of (1.1) and has a meaning for functions u, $[0,\infty) \to D$, such that $u \in C^1([0,\infty); H)$ (no need therefore of a second derivative u'' !). (Note that $v_0 = u'(0)$.)

We define now, as usual, the weak solution of (1.1)-(1.2) if (1.9) holds $\forall \varphi(\cdot)$ in the above class of test-functions, and if, furthermore, $u(0) = u_0$.

Let us define also, associated to any such weak solution, the 'energy' $E(t)$ $(t \geq 0)$ by the formula

$$E(t) = \frac{1}{2}\|u'(t)\|^2 - \frac{1}{2}\mathcal{M}(u(t),u(t)) \tag{1.10}$$

and we shall assume the 'weak conservation principle':

$$E(t) \le E(0)\ , \quad \forall t \ge 0\ . \tag{1.11}$$

We are now ready for the main statement:

Theorem 1.1 *Let f, $[0,\infty) \to (0,\infty)$ be a C^2-function, such that $(\ln f)'' \ge 0$. Let u be a weak solution - as in (1.9) - such that (1.10) holds too. Then*

$$(u(t),u'(t)) \ge -\frac{f'(t)}{f(t)}\|u(t)\|^2 + \frac{1}{f^2(t)}\left\{f(0)f'(0)\|u_0\|^2 + f^2(0)(u_0,v_0)\right\}$$
$$-2\frac{E(0)}{f^2(t)}\int_0^t f^2(\eta)\,\mathrm{d}\eta\ . \tag{1.12}$$

Proof Let us define the function $w(\cdot)$, $[0,\infty) \to H$, by the formula

$$w(t) = f(t)u(t)\ , \quad \forall t \ge 0\ , \tag{1.13}$$

and then

$$\varphi(t) = f(t)w(t) = f^2(t)u(t)\ . \tag{1.14}$$

Note first that $\varphi(\cdot)$ is a test-function ($\varphi(\cdot) \in C^1([0,\infty);H)$ and $\varphi(t) \in D\ \forall t \ge 0$).

We next introduce $\varphi(\cdot)$ and $u(\cdot)$ in (1.9). We see that $u(t) = \dfrac{1}{f(t)}w(t)$, $\forall t \ge 0$, hence

$$u'(t) = \frac{1}{f(t)}w'(t) - \frac{f'(t)}{f^2(t)}w(t) = \frac{1}{f(t)}\left[w'(t) - \frac{f'(t)}{f(t)}w(t)\right]$$
$$= \frac{1}{f(t)}\left[w'(t) - (\ln f(t))w(t)\right],\quad t \ge 0\ . \tag{1.15}$$

Also, we compute the derivative of $\varphi(\cdot)$; one obtains

$$\varphi'(t) = f'(t)w(t) + f(t)w'(t) = f(t)\left(w'(t) + \frac{f'(t)}{f(t)}w(t)\right)$$
$$= f(t)(w'(t) + (\ln f(t))'w(t))\ . \tag{1.16}$$

We substitute therefore in (1.9) and obtain

$$(\varphi(t),u'(t))=\left(f(t)w(t),\frac{1}{f(t)}[w'(t)-(\ln f(t))w(t)]\right)=(w(t),w'(t))$$
$$-(w(t),(\ln f)'(t)w(t))=(w(t),w'(t)-(\ln f)'(t)w(t)) \tag{1.17}$$

$$(\varphi(0),v_0)=\left(f^2(0)u(0),v_0\right)=f^2(0)\left(u_0,v_0\right) \quad \text{(from (1.2))} \tag{1.18}$$

$$(\varphi'(t),u'(t))=\left(f(t)[w'(t)+(\ln f)'(t)w(t)],\frac{1}{f(t)}[w'(t)-(\ln f)'(t)w(t)]\right)$$
$$=(w'(t)+(\ln f)'(t)w(t),w'(t)-(\ln f)'(t)w(t))$$
$$=\|w'(t)\|^2-[(\ln f)'(t)]^2\|w(t)\|^2 \tag{1.19}$$

$$\mathcal{M}(\varphi(\cdot),u(\cdot))=\mathcal{M}\left(fw,\frac{1}{f}w\right)=\mathcal{M}(w(\cdot),w(\cdot))\ . \tag{1.20}$$

Putting these equalities in (1.9) we come to the relation:

$$(w,w'-(\ln f)'w)=f^2(0)\left(u_0,v_0\right)+\int_0^t\left\{\|w'(\eta)\|^2-[(\ln f)'(\eta)]^2\|w(\eta)\|^2\right\}\mathrm{d}\eta$$
$$+\int_0^t\mathcal{M}(w(\eta),w(\eta))\,\mathrm{d}\eta\ . \tag{1.21}$$

At this stage we shall make use of the 'weak conservation principle' (1.11) which can be written as:

$$\tfrac{1}{2}(u'(t),u'(t))-\tfrac{1}{2}\mathcal{M}(u(t),u(t))\le E(0)\ , \qquad \forall t\ge 0\ . \tag{1.22}$$

As previously seen (in 1.15), we have, $\forall t\ge 0$,

$$u(t)=\frac{1}{f(t)}w(t)\ , \qquad u'(t)=\frac{1}{f(t)}[w'(t)-(\ln f(t))'w(t)]$$

whence, introducing in (1.22), one obtains

$$\tfrac{1}{2}\left(\frac{1}{f}[w'-(\ln f)'w],\ \frac{1}{f}[w'-(\ln f)'w]\right)-\tfrac{1}{2}\mathcal{M}\left(\frac{1}{f}w,\ \frac{1}{f}w\right)\le E(0)\ ;$$

consequently, one gets:

$$\frac{1}{f^2(t)}\|w'(t)\|^2+\frac{1}{f^2(t)}[(\ln f)'(t)]^2\|w(t)\|^2-2\frac{(\ln f)'(t)}{f^2(t)}(w,w')(t)$$
$$-\ \frac{1}{f^2(t)}\mathcal{M}(w(t),w(t))\le 2E(0) \tag{1.23}$$

which in turn becomes

$$\|w'(t)\|^2+[(\ln f)'(t)]^2\|w(t)\|^2-2(\ln f)'(t)(w(t),w'(t))-\mathcal{M}(w(t),w(t))$$
$$\le 2f^2(t)E(0)$$

that is also

$$\mathcal{M}(w(t),w(t))\ge -2f^2(t)E(0)+\|w'(t)\|^2$$
$$-2(\ln f)'(t)(w(t),w'(t))+[(\ln f)'(t)]^2\|w(t)\|^2 \tag{1.24}$$

which, when introduced in the equality (1.21), implies the lower estimate:

$$(w'(t)-(\ln f)'(t)w(t),w(t))\ge f^2(0)(u_0,v_0)+\int_0^t\left\{\|w'(\eta)\|^2-[(\ln f)'(\eta)]^2\|w(\eta)\|^2\right\}\mathrm{d}\eta$$
$$-2E(0)\int_0^t f^2(\eta)\,\mathrm{d}\eta+\int_0^t\|w'(\eta)\|^2\,\mathrm{d}\eta-2\int_0^t(\ln f)'(\eta)(w(\eta),w'(\eta))\,\mathrm{d}\eta \tag{1.25}$$
$$+\int_0^t[(\ln f)'(\eta)]^2\|w(\eta)\|^2\,\mathrm{d}\eta\ .$$

We next transform, using integration by parts, to get

$$\int_0^t(\ln f)'(\eta)(w(\eta),w'(\eta))\,\mathrm{d}\eta=(\ln f)'(t)\tfrac{1}{2}\|w(t)\|^2-(\ln f)'(0)\tfrac{1}{2}\|w(0)\|^2$$
$$-\tfrac{1}{2}\int_0^t\|w(\eta)\|^2(\ln f)''(\eta)\,\mathrm{d}\eta \tag{1.26}$$

whence, substituting in (1.25)

$$(w'(t)-(\ln f)'(t)w(t),w(t)) \geq f^2(0)(u_0,v_0)+2\int_0^t \|w'(\eta)\|^2\, d\eta - 2E(0)\int_0^t f^2(\eta)\, d\eta$$

$$-(\ln f)'(t)\|w(t)\|^2 + \frac{f'(0)}{f(0)}\|w(0)\|^2 + \int_0^t \|w(\eta)\|^2 (\ln f)''(\eta)\, d\eta$$

$$\geq f^2(0)(u_0,v_0)+\int_0^t \|w(\eta)\|^2 (\ln f)''(\eta)\, d\eta$$

$$-2E(0)\int_0^t f^2(\eta)\, d\eta - (\ln f)'(t)\|w(t)\|^2 + \frac{f'(0)}{f(0)}\|w(0)\|^2 . \tag{1.27}$$

Let us make use of (1.13). Then $w'(t)=f'(t)u(t)+f(t)u'(t)$ and

$$(w'(t),w(t))=(f'(t)u(t)+f(t)u'(t),\, f(t)u(t)) = f'(t)f(t)\|u(t)\|^2$$

$$+f^2(t)(u'(t),u(t)) . \tag{1.28}$$

Also, we see that

$$((\ln f)'(t)w(t),w(t)) = \left(\frac{f'(t)}{f(t)}f(t)u(t),\, f(t)u(t)\right) = (f'(t)u(t),\, f(t)u(t))$$

$$= f'(t)f(t)\|u(t)\|^2 ; \tag{1.29}$$

consequently one obtains that

$$(w'(t)-(\ln f)'(t)w(t),w(t)) = f'(t)f(t)\|u(t)\|^2 + f^2(t)(u'(t),u(t))$$

$$-f'(t)f(t)\|u(t)\|^2 = f^2(t)(u'(t),u(t)) . \tag{1.30}$$

Introducing this last equality into the estimate (1.27) we conclude that

$$f^2(t)(u'(t),u(t)) \geq f^2(0)(u_0,v_0)+\int_0^t \|w(\eta)\|^2(\ln f)''(\eta)\, d\eta$$

$$-2E(0)\int_0^t f^2(\eta)\, d\eta - (\ln f)'(t)f^2(t)\|u(t)\|^2 + \frac{f'(0)}{f(0)}f^2(0)\|u(0)\|^2$$

or also

$$(u(t),u'(t)) \geq -\frac{f'(t)}{f(t)}\|u(t)\|^2 - 2\frac{E(0)}{f^2(t)}\int_0^t f^2(\eta)\,\mathrm{d}\eta + \frac{1}{f^2(t)} f(0)f'(0)\|u_0\|^2$$

$$+\frac{1}{f^2(t)} f^2(0)(u_0, v_0), \qquad \forall t \geq 0 \tag{1.31}$$

(we use the assumption $(\ln f)'' \geq 0$). This is exactly (1.12), hence Theorem 1.1 is proved. ■

2. The estimate (1.12) entails the following uniqueness result for the Cauchy problem:

Theorem 2.1 *Let* $u(\cdot) \in C^1([0,\infty); H)$ *such that* $u(t) \in D$, $\forall t \geq 0$, $u(0) = u'(0) = \mathbf{0}$, *and*

$$(\varphi(t), u'(t)) = \int_0^t [(\varphi'(\eta), u'(\eta)) + \mathcal{M}(\varphi(\eta), u(\eta))]\,\mathrm{d}\eta$$

$\forall \varphi(\cdot) \in C^1([0,\infty); H)$, such that $\varphi(t) \in D$, $\forall t \geq 0$, and assume also that $E(t) \leq E(0)$.

Then $u(t) = \mathbf{0}$, $\forall t \geq 0$.

Proof We apply Theorem 1.1 taking functions $f_\lambda(t) = \mathrm{e}^{-\lambda t}$, $\lambda \geq 0$. If, as we assume, $u(0) = u'(0) = \mathbf{0}$, it follows that $E(0) = 0$ (from (1.10)). The inequality (1.12) now becomes

$$(u(t), u'(t)) \geq \lambda \|u(t)\|^2,$$

that is

$$\|u(t)\|^2 \leq \frac{1}{\lambda}(u(t), u'(t)), \qquad \forall t \geq 0$$

and, $\forall \lambda > 0$.

As $\lambda \to \infty$ we obtain

$$u(t) = \mathbf{0}, \qquad \forall t \geq 0.$$

■

Chapter X

Uniqueness of ultraweak solutions for a first-order abstract Cauchy problem in Banach space

Introduction

In this chapter we extend to 'ultraweak solutions' (conveniently defined) a simple result about uniqueness of the abstract Cauchy problem (Hille-Phillips [5], Theorem 23.7.1) for the equation $u'(t) = Au(t),\ 0 \le t < \infty$, depending on the eigenvalues of the operator A.

1. The underlying Banach space X is assumed 'reflexive', that is $J(X) = X''$ (this restriction does not appear in [5]) (see Chapter III).

Then, consider a linear closed operator A, $\mathcal{D}(A) \subset X \to X$, and let A' be its dual operator (see Chapter II-15). Thus, we also assume that $\mathcal{D}(A)$ is dense in X. Our aim here, first of all, is to establish that the domain $\mathcal{D}(A')$ is also dense in the dual space X'.

Let us start with:

Proposition 1.1 *Let T be a linear operator, $\mathcal{D}(T) \subset E \to F$, where E,F are Banach spaces. Assume T is closable. It follows that $\forall y \in F,\ y \neq \mathbf{0}$, the element $(\mathbf{0}, y)$ in $E \times F$ does not belong to $\overline{G_T}$.*

Remark We refer here to Chapter II-9 (the graph $G(T)$ is here denoted by G_T).

Proof As T is closable, $\exists$ closed linear operator $\tilde{T}$, $\mathcal{D}(\tilde{T}) \subset E \to F$, such that $T \subset \tilde{T}$. Then $G_T \subset G_{\tilde{T}}$ and $G_{\tilde{T}}$ is closed in $E \times F$. Then the closure $\overline{G_T}$ is contained in $G_{\tilde{T}}$.

Now, if $(\mathbf{0}, y) \in \overline{G_T}$, it follows that $(\mathbf{0}, y) \in G_{\tilde{T}}$, hence $y = \tilde{T}\mathbf{0} = \mathbf{0}$. ■

Next we have:

Proposition 1.2 *Let T be a linear operator, $\mathcal{D}(T) \subset E \to F$, where E,F are Banach spaces. Assume that $\forall y \in F,\ y \neq \mathbf{0}$, the element $(\mathbf{0}, y) \notin \overline{G_T}$. Then T has a minimal closed linear extension.*

Remark See Theorem 9.1, Chapter II.

Proof Let us define an operator $S = \overline{T}$ in the following way:

$$\mathcal{D}(S) = \{x \in E, \text{ such that } \exists z \in F \text{ with } (x,z) \in \overline{G_T}\} .$$

For $x \in \mathcal{D}(S)$, $Sx = z$.

Now, S is a single-valued operator: assume in fact that $z_1, z_2 \in F$ and $(x, z_1) \in \overline{G}_T$, $(x, z_2) \in \overline{G}_T$. Then $(x, z_1) - (x, z_2) = (\mathbf{0}, z_1 - z_2) \in \overline{G}_T$ (which is a linear closed subspace of $E \times F$).

Thus, from our assumption $z_1 - z_2 = \mathbf{0}$, $z_1 = z_2$.

Next, S is a linear operator: let $x_1, x_2 \in \mathcal{D}(S)$; then, $\exists! z_1$, $\exists! z_2$ in E, with $(x_1, z_1) \in \overline{G}_T$, $(x_2, z_2) \in \overline{G}_T$. Therefore $(x_1 + x_2, z_1 + z_2) \in G_T$, where $z_1 + z_2 \in F$. Accordingly: $x_1 + x_2 \in \mathcal{D}(S)$ and $S(x_1 + x_2) = z_1 + z_2 = Sx_1 + Sx_2$.

Also, for $x \in \mathcal{D}(S)$, $\lambda \in K$ (the field of E,F), $\exists! z \in F$, $(x,z) \in \overline{G}_T$ hence $(\lambda x, \lambda z) = \lambda(x,z) \in \overline{G}_T$, and $\lambda x \in \mathcal{D}(S)$. Furthermore,

$$S(\lambda x) = \lambda z = \lambda Sx .$$

Now, S is also a closed operator: in fact $G_S = \{(x,z), x \in \mathcal{D}(S)\} = \{(x, Sx), x \in \mathcal{D}(S)\}$. Take $x_n \in \mathcal{D}(S)$, $x_n \to x_0$ in E, and $Sx_n \to y_0$ in F. Then, $\exists! z_n \in F$ $(\forall n \in \mathbb{N})$, such that $(x_n, z_n) \in \overline{G}_T$ and $Sx_n = z_n$. Thus, $(x_n, z_n) \to (x_0, y_0)$ which belongs again to $\overline{G}_T$, and therefore $x_0 \in \mathcal{D}(S)$, $y_0 = Sx_0$.

Next, S is a (closed) extension of T. In fact, if $x \in \mathcal{D}(T)$, take $z = Tx$, and $(x,z) = (x, Tx) \in G_T \subset \overline{G}_T$. Therefore $z = Sx$ too.

Finally, S is a minimal closed linear extension of T. In fact, let T_1 be a linear operator on $\mathcal{D}(T_1) \subset E$ to F, T_1 closed and $T_1 \supset T$. It follows that $G_{T_1} \supset G_T$ hence $G_{T_1} \supset \overline{G}_T$ and $\overline{G}_T \supset G_S$ hence $G_{T_1} \supset G_S$, $T_1 \supset S$. ■

Before proceeding we shall give the important definition of a 'total' set of linear functionals.

Definition 1.1 *Let V be a vector space and $\mathcal{F}$ be a set of linear functionals on V. Then $\mathcal{F}$ is a total set if, given any $v \in V$, $v \neq \mathbf{0}$, there exists $f \in \mathcal{F}$ such that $f(v) \neq 0$.*

We shall establish:

Proposition 1.3 *Let T be a linear operator with dense domain $\mathcal{D}(T) \subset E \to F$ and T' be its dual operator: $\mathcal{D}(T') \subset F' \to E'$. Then $\mathcal{D}(T')$ is total if and only if, $\forall y \in F$, $y \neq \mathbf{0}$, $(\mathbf{0}, y) \notin \overline{G_T}$.*

Proof

(1) Let us assume that $\mathcal{D}(T')$ is total and let us assume also that for some $y \in F$, we have $(\mathbf{0}, y) \in \overline{G_T}$. It follows accordingly that there exists a sequence (x_n) where $x_n \in \mathcal{D}(T)$, $\forall n \in \mathbb{N}$, $x_n \to \mathbf{0}$ in E, $Tx_n \to y$ in F. Take now $y' \in \mathcal{D}(T')$ (thus $y' \in F'$) and write the equality (resulting from the definition of T'):

$$y'(Tx_n) = (T'y')(x_n) \ . \tag{1.1}$$

As $x_n \to \mathbf{0}$ in E and $T'y' \in E'$, we find that $\lim_{n\to\infty} (T'y')(x_n) = 0$. On the other hand, as $Tx_n \to y$ in F and $y' \in F'$, we find that

$$y'(Tx_n) \to y'(y) \ .$$

Thus $y'(y) = 0$, $\forall y' \in \mathcal{D}(T')$. As $\mathcal{D}(T')$ is total, it follows that $y = \mathbf{0}$.

(2) Let us assume now that $y \in F$, $y \neq \mathbf{0}$, entails that $(\mathbf{0}, y) \notin \overline{G_T}$. We shall make use of a classical corollary of the Hahn-Banach extension theorem (see for instance [5], Theorem 2.75, p. 30), and obtain the existence of an element $z' \in (E \times F)'$, such that $z'(z) = 0$, $\forall z \in \overline{G_T}$ and $z'((\mathbf{0}, y)) \neq 0$. Define now elements $x' \in E'$ and $y' \in F'$ by $x'(x) = z'((x, \mathbf{0}))$, $\forall x \in E$, $y'(y) = z'((\mathbf{0}, y))$, $\forall y \in F'$. It is easy to prove that in fact $x' \in E'$ and $y' \in F'$. It follows that, $\forall x \in \mathcal{D}(T)$

$$z'((x, Tx)) = z'((x, \mathbf{0})) + z'((\mathbf{0}, Tx)) = x'(x) + y'(Tx) = 0$$

as $z'(z) = 0$, $\forall z \in \overline{G_T}$.

This equality shows that $y' \in \mathcal{D}(T')$ and $(T'y')(x) = -x'(x)$, $\forall x \in \mathcal{D}(T)$.
We also have $z'((\mathbf{0}, y)) \neq 0$, that is, $y'(y) \neq 0$.

We found therefore, if $(\mathbf{0}, y) \notin \overline{G_T}$ and $y \neq \mathbf{0}$, that these exists $y' \in \mathcal{D}(T')$ such that $y'(y) \neq 0$. Thus $\mathcal{D}(T')$ is total (as a subset of F'). ■

From Proposition 1.1-1.2 and 1.3 we derive:

Corollary 1.1 *Let T, $\mathcal{D}(T) \subset E \to F$ be a linear operator with dense domain; then $\mathcal{D}(T')$ is total if and only if T is a closable operator.*

Before proceeding further let us remember (Chapter II-9) that if the linear operator T is closable (i.e., it has a closed extension), then it also has a minimal closed extension $\overline{T}$ (its 'closure'). We now have:

Proposition 1.4 *Let T, $\mathcal{D}(T) \subset E \to F$ be a linear operator with dense domain which is closable. Then $T' = (\overline{T})'$.*

Proof The proof consists on three parts.

(i) $\mathcal{D}((\overline{T})') \subset \mathcal{D}(T')$.

Let us remember also (Chapter II-15) that if A, $\mathcal{D}(A) \subset E \to F$ is a linear operator with dense domain, the domain $\mathcal{D}(A') = \mathcal{D}'$ of its dual operator is composed of those $f' \in F'$ such that $f' \circ A$ is continuous on $\mathcal{D}(A)$ (as a linear mapping: $\mathcal{D}(A) \to K$ - the field of E,F). Also, the equality $(f' \circ A)(e) = f'(Ae) = (A'f')(e)$, $\forall e \in D(A)$, holds true.

Take $y' \in \mathcal{D}((\overline{T})')$ (hence, $y' \in F'$). Then $y' \circ \overline{T}$ is (linear) continuous, $\mathcal{D}(\overline{T}) \to K$, hence, as $T \subset \overline{T}$, it is also linear continuous on $\mathcal{D}(T)$. Thus $y' \in \mathcal{D}(T')$.

(ii) $\mathcal{D}(T') \subset \mathcal{D}((\overline{T})')$. Take $y' \in \mathcal{D}(T')$. We are to prove that $y' \circ \overline{T}$ is continuous on $\mathcal{D}(\overline{T})$.

Take $x \in \mathcal{D}(\overline{T})$ ($x \in E$). Then the element $(x, \overline{T}x)$ in $E \times F$ actually belongs to $G_{\overline{T}}$ which equals $\overline{G_T}$ (see again Chapter II-9). There exists accordingly a sequence (x_n) of elements in $\mathcal{D}(T)$, such that $x_n \to x$ (in E) while $Tx_n \to \overline{T}x$ (in F). Then

$$y' \circ \overline{T}x = \lim_{n \to \infty} y' \circ \overline{T}x_n ,$$

and

$$\begin{aligned}|(y' \circ \overline{T})(x)| &= \lim |(y' \circ \overline{T})(x_n)| \\ &= \lim |(y' \circ T)(x_n)| = \lim |(T'y')(x_n)| \le \|(T'y')\| \lim \|x_n\| = \|(T)'y'\| \, \|x\|\end{aligned}$$

which means precisely that $y' \circ \overline{T}$ is continuous on $\mathcal{D}(\overline{T})$.

(iii) Thus, we now have $\mathcal{D}((\overline{T})') = \mathcal{D}(T')$. Take $y' \in \mathcal{D}(T')$; we obtain, $\forall x \in \mathcal{D}(T)$

$$(T'y')(x) = y'(Tx) = y'(\overline{T}x) = ((\overline{T})'y')(x) \ .$$

Thus, the elements in E', $T'y'$ and $(\overline{T})'y'$ are equal on the dense set $\mathcal{D}(T)$, hence everywhere on E. We proved $T' = (\overline{T})'$. ■

Corollary 1.2 *Let T, $\mathcal{D}(T) \subset E \to F$, be a linear operator with dense domain; then '$\mathcal{D}(T')$ is total' implies $T' = (\overline{T})'$.*

Consider now the special case of a reflexive Banach space. First note:

Proposition 1.5 *Let X be a reflexive B-space and $\mathcal{A}$ be linear subspace of X'. Then $\mathcal{A}$ is total if and only if it is dense in X'.*

Proof

(i) First, let us assume that $\mathcal{A}$ is dense in X'. Let $x \in X$ and $x'(x) = 0$, $\forall x' \in \mathcal{A}$. Then, obviously, $x'(x) = 0$, $\forall x' \in X'$, and (as seen in Chapter III) it follows that $x = \mathbf{0}$. Therefore, $\mathcal{A}$ is a total set.

(ii) Let us assume that $\mathcal{A}$ is a total set but $\mathcal{A}$ is *not* dense in X'. Accordingly, its closure $\overline{\mathcal{A}}$ in X' is a strict (proper) subspace of X'. Again we use Theorem 2.7.5, p. 30 of [5] to obtain the existence of an element x'' of the (second dual) space X'', with the property that

$$x''(x') = 0 \qquad \forall x' \in \mathcal{A},\ x'' \neq \mathbf{0} \ .$$

Now, the reflexivity of X means that $X'' = \mathcal{J}(X)$; therefore, $\forall x'' \in X''$, there exists $x \in X$, such that

$$x''(x') = x'(x) \ , \qquad \forall x' \in X' \qquad (\text{and } x'' = \mathcal{J}x) \ .$$

Also, we know that $\|\mathcal{J}x\| = \|x''\| = \|x\|$, hence, if $x'' \neq \mathbf{0}$, then $x \neq \mathbf{0}$.

On the other hand, if $x' \in \mathcal{A}$, we have $x'(x) = x''(x') = 0$. As $\mathcal{A}$ is total, this should imply that $x = \mathbf{0}$, and we have a contradiction. Therefore, the total set $\mathcal{A}$ *is* dense in X'. ■

Corollary 1.3 *Let T, $\mathcal{D}(T) \subset E \to F$ be a linear operator with dense domain, and F be a reflexive B-space. Then T is closable if and only if $\mathcal{D}(T')$ is dense in F'.*

(This is immediate from Corollary 1.1 and Proposition 1.5.)

Thus, returning to the differential equation $u' = Au$ with linear closed operator A, $\mathcal{D}(A) \subset X \to X$, which is *reflexive*, we see that indeed, when $\overline{\mathcal{D}(A)} = X$ then also $\overline{\mathcal{D}(A')} = X'$.

2. Our next goal is to give formal definitions and some simple properties of the concept of ultraweak solutions; this definition is akin to the well-known concept of a 'solution in the sense of distributions' for partial differential equations. As in that theory, we need classes of test-functions, which are chosen according to the problem under investigation. In particular, in our present situation the test-functions from the class denoted by $K_{A'}(0,\infty) = \{\varphi'(t) \in C_0^1(]0,\infty[;X);\ \varphi'(t) \in \mathcal{D}(A')\forall t \geq 0,\ (A'\varphi')(\cdot) \in C([0,\infty[;X')\}$.

Thus, the functions in $K_{A'}(0,\infty)$ are vector-valued functions defined on $[0,\infty)$, with range in the dual space X', once continuously differentiable and vanishing outside some compact interval $[a,b]$ contained in $(0,\infty)$ - depending on the function; furthermore, their range is contained in the domain $\mathcal{D}(A')$ and $(A'\varphi')(\cdot)$ is continuous, $[0,\infty) \to X'$.

(Do not confuse the notation of φ', which here means a function with range in X' with the derivative $\frac{\mathrm{d}}{\mathrm{d}t}\varphi'$ which will appear shortly.)

Now, the continuous function $u(t)$, $[0,\infty) \to X$, is said to be an ultraweak solution of the equation

$$u'(t) = Au(t)\ , \qquad 0 \leq t < \infty \tag{2.1}$$

if

$$\int_0^\infty \left(\frac{\mathrm{d}}{\mathrm{d}t}\varphi'(t) + (A'\varphi')(t)\right)(u(t))\,\mathrm{d}t = 0 \tag{2.2}$$

holds true, $\forall \varphi' \in K_{A'}(0,\infty)$.

(Note that the function $\frac{\mathrm{d}}{\mathrm{d}t}\varphi' + A'\varphi'$ is continuous, $[0,\infty) \to X'$, while $u(\cdot)$ is continuous, $[0,\infty) \to X$; it follows readily that the integrand in (2.2) is a continuous function, $[0,\infty) \to K$ so that the integral (2.2) makes sense.)

Precisely, let $t_n \in [0,\infty)$, $t_n \to t_0 \in [0,\infty)$; then, if $\psi' = \frac{\mathrm{d}}{\mathrm{d}t}\varphi' + A'\varphi'$, we have $\psi'(t_n)(u(t_n)) - \psi'(t_0)(u(t_0)) = [\psi'(t_n) - \psi'(t_0)](u(t_n)) + \psi'(t_0)(u(t_n) - u(t_0))$ and

$$\left|\psi'(t_n)(u(t_n)) - \psi'(t_0)(u(t_0))\right| \le \left\|\psi'(t_n) - \psi'(t_0)\right\|_{X'} \left\|u(t_n)\right\|_X$$
$$+ \left\|\psi'(t_0)\right\|_{X'} \left\|u(t_n) - u(t_0)\right\| \to 0 \quad \text{as } n \to \infty .$$

We shall prove in this chapter the following:

Theorem 2.1 *Let X be a reflexive B-space and let A,* $\mathcal{D}(A) \subset X \to X$ *be a linear closed operator with dense domain, such that the set of its eigenvalues is not dense in any half-plane* $\{\lambda \in \mathbb{C},\ \operatorname{Re}\lambda > \alpha\}$. *Consider the continuous function* $u(\cdot)$, $[0,\infty) \to X$, *satisfying (2.2) and also*

$$u(0) = \mathbf{0}\,, \qquad \limsup_{t\to\infty} \frac{1}{t}\log\|u(t)\| = w < +\infty\,. \tag{2.3}$$

Then $u(t) = \mathbf{0}$, $\forall t \ge 0$.

Proof

(a) To start the proof, note first that

$$w = \lim_{R\to\infty}\left(\sup_{t\ge R} \frac{1}{t}\log\|u(t)\|\right),$$

hence $\forall w_1 > w$, we have

$$\sup_{t\ge R} \frac{1}{t}\log\|u(t)\| < w_1$$

when $R \ge R_1$ and therefore,

$$\frac{1}{t}\log\|u(t)\| < w_1\,, \qquad \forall t \ge R \ge R_1\,;$$

thus

$$\frac{1}{t}\log\|u(t)\| < w_1, \quad \forall t \geq R_1, \quad \text{and} \quad \|u(t)\| \leq \exp(tw_1)$$

for all $t \geq R_1$.

Next, one considers a sequence $\{\alpha_n(\cdot)\}$ of scalar-valued, non-negative C^1-functions, vanishing for $|t| > \frac{1}{n}$, such that $\int_{\mathbb{R}} \alpha_n(s)\,\mathrm{d}s = 1$, $\forall n = 1,2,\ldots$. This enables us to calculate convolutions $(u * \alpha_n)(\cdot)$ given by

$$(u * \alpha_n)(t) = \int_{\mathbb{R}} u(s)\alpha_n(t-s)\,\mathrm{d}s = \int_{t-\frac{1}{n}}^{t+\frac{1}{n}} u(s)\alpha_n(t-s)\,\mathrm{d}s\,. \tag{2.4}$$

Precisely, as $u(\cdot)$ is only defined on $[0,\infty)$, $(u * \alpha_n)(\cdot)$ will be defined on $[t_0,\infty)$ where $t_0 > 0$, $t_0 - \frac{1}{n_0} \geq 0$ and $n \geq n_0$.

We now obtain an estimate for $\|u * \alpha_n\|$. Use the formula

$$(u * \alpha_n)(t) = \int_{-\frac{1}{n}}^{\frac{1}{n}} u(t-\sigma)\alpha_n(\sigma)\,\mathrm{d}\sigma$$

and derive, first of all, that

$$\|(u * \alpha_n)(t)\| \leq \int_{-\frac{1}{n}}^{\frac{1}{n}} \|u(t-\sigma)\|\alpha_n(\sigma)\,\mathrm{d}\sigma, \quad t \geq t_0 \text{ and } n \geq n_0\,. \tag{2.5}$$

Assume, as a special case, that $t \geq R_1 + 1$ and $-\frac{1}{n} < \sigma < \frac{1}{n}$; then $t - \sigma \geq t - \frac{1}{n} \geq t - 1 \geq R_1$ and accordingly: $\|u(t-\sigma)\| \leq \mathrm{e}^{(t-\sigma)w_1}$ so that (2.5) now gives the estimate

$$\|(u*\alpha_n)(t)\| \le \int_{-\frac{1}{n}}^{\frac{1}{n}} \mathrm{e}^{(t-\sigma)w_1}\alpha_n(\sigma)\,\mathrm{d}\sigma$$

$$= \mathrm{e}^{tw_1}\int_{-\frac{1}{n}}^{\frac{1}{n}} \mathrm{e}^{-\sigma w_1}\alpha_n(\sigma)\,\mathrm{d}\sigma$$

$$\le \mathrm{e}^{tw_1} \sup_{-1\le\sigma\le 1}\left(\mathrm{e}^{-\sigma w_1}\right) = C_{w_1}\mathrm{e}^{tw_1}\,, \qquad \forall t \ge R_1+1,\ \forall n\in\mathbb{N}\,. \tag{2.6}$$

Consider now the improper Riemann integral

$$\int_{t_0}^{\infty} \mathrm{e}^{-\lambda t}(u*\alpha_n)(t)\,\mathrm{d}t\,, \qquad \text{for } t_0 \ge \frac{1}{n_0} > 0 \text{ and } n \ge n_0\,.$$

This integral is absolutely convergent, $\forall\lambda\in\mathbb{C},\ \mathrm{Re}\,\lambda > w$. In fact, once such a number λ is given, we choose w_1 where $w < w_1 < \mathrm{Re}\,\lambda$. The expression under the integral sign is then estimated in norm by:

$$\left\|\mathrm{e}^{-\lambda t}(u*\alpha_n)(t)\right\| \le \mathrm{e}^{-(\mathrm{Re}\,\lambda)t}\|(u*\alpha_n)(t)\| \le \mathrm{e}^{-(\mathrm{Re}\,\lambda)t}\cdot C_{w_1}\mathrm{e}^{tw_1} \tag{2.7}$$

for $t \ge R_1+1$, $n\ge n_0$ hence, for $t\ge R_1+1$, we get

$$\left\|\mathrm{e}^{-\lambda t}(u*\alpha_n)(t)\right\| \le C_{w_1}\mathrm{e}^{t(w_1-\mathrm{Re}\,\lambda)}\,;$$

as

$$\int_{t_0}^{\infty} \mathrm{e}^{t(w_1-\mathrm{Re}\,\lambda)}\,\mathrm{d}t < \infty$$

we have absolute convergence.

(b) In this section of the proof we essentially show that if $u(\cdot)$ is a solution to (2.2) then $(u*\alpha_n)(\cdot)$ is a solution to (2.1) in the ordinary sense.

We follow, say, the pattern appearing in [21-II], Theorem 3.1, p. 80.

Take $t_0 > 0$, $\frac{1}{n_0} \le t_0$ and $n > n_0$. Then, for $t > t_0$, consider functions

$$\varphi'_{t,n}(\tau) = \alpha_n(t-\tau)\phi' , \quad \text{where } \phi' \in \mathcal{D}(A'),\ n \geq n_0 . \tag{2.8}$$

We see that all functions in (2.8) belong to $K_{A'}(0,\infty)$: they are well defined for $\tau \in (0,\infty)$, have range in $\mathcal{D}(A')$, hence in X', are as regular as $\alpha_n(\cdot)$, and hence belong to $C^1(0,\infty;X')$; furthermore, $\varphi'_{t,n}(\tau) = \mathbf{0}$ if $|t-\tau| > \frac{1}{n}$, that is for $\tau \notin \left[t - \frac{1}{n},\ t + \frac{1}{n}\right]$. Now, as $t > t_0$, we have $t - \frac{1}{n} > t_0 - \frac{1}{n} \geq t_0 - \frac{1}{n_0} \geq 0$, $t - \frac{1}{n} > 0$. Hence $\varphi'_{t,n}(\tau) = \mathbf{0}$ outside the compact interval $\left[t - \frac{1}{n},\ t + \frac{1}{n}\right]$ which is contained in $(0,+\infty)$.

We are thus entitled to introduce these functions $\varphi'_{t,n}(\tau)$ as test-functions in the above equality (2.2). Note that:

$$\frac{\mathrm{d}}{\mathrm{d}\tau}\varphi'_{t,n}(\tau) = -\dot{\alpha}_n(t-\tau)\phi'$$

(where the dot $\cdot$ again means derivative).

We obtain accordingly, from (2.2), the relation

$$\int_0^\infty \left(-\dot{\alpha}_n(t-\tau)\phi' + \alpha_n(t-\tau)A'\phi'\right)(u(\tau))\,\mathrm{d}\tau = 0 \tag{2.9}$$

which can be written as

$$(A'\phi')\left(\int_0^\infty \alpha_n(t-\tau)u(\tau)\,\mathrm{d}\tau\right) = \phi'\left(\int_0^\infty \dot{\alpha}_n(t-\tau)u(\tau)\,\mathrm{d}\tau\right), \quad \forall \phi' \in \mathcal{D}(A') \quad . \tag{2.10}$$

Note also that

$$\int_0^\infty \alpha_n(t-\tau)u(\tau)\,\mathrm{d}\tau = \int_{-\infty}^\infty \alpha_n(t-\tau)u(\tau)\,\mathrm{d}\tau \quad \text{for } n > n_0, t > t_0$$

(for $\alpha_n(t-\tau) = 0$ for $|t-\tau| \geq \frac{1}{n}$, hence for $\tau \notin \left[t - \frac{1}{n},\ t + \frac{1}{n}\right]$ where, as previously seen, $\left[t - \frac{1}{n},\ t + \frac{1}{n}\right] \subset (0,+\infty)$.

Thus, $\int_0^\infty \alpha_n(t-\tau)u(\tau)\,\mathrm{d}\tau = (u*\alpha_n)(t)$ for $t>t_0$, $n>n_0$, and introducing also the canonical imbedding $\mathcal{J}$ from X into X'' we can write (2.10) in the form

$$\left(\mathcal{J}(u*\alpha_n)(t)\right)(A'\phi') = \mathcal{J}\left(\frac{\mathrm{d}}{\mathrm{d}t}(u*\alpha_n)\right)(\phi'), \quad \forall\phi'\in\mathcal{D}(A') \tag{2.11}$$

(we use here also the relation

$$\frac{\mathrm{d}}{\mathrm{d}t}(u*\alpha_n)(t) = \frac{\mathrm{d}}{\mathrm{d}t}\int_0^\infty \alpha_n(t-\tau)u(\tau)\,\mathrm{d}\tau = \int_0^\infty \dot{\alpha}_n(t-\tau)u(\tau)\,\mathrm{d}\tau ,$$

which is immediate).

Using the definition of the second dual operator A'' we can interpret (2.11) in the form: (one uses (15.6) - Chapter II) $\mathcal{J}(u*\alpha_n)(t)\in\mathcal{D}(A'')$ and

$$A''\left(\mathcal{J}(u*\alpha_n)\right)(t) = \mathcal{J}\left(\frac{\mathrm{d}}{\mathrm{d}t}(u*\alpha_n)(t)\right) = \frac{\mathrm{d}}{\mathrm{d}t}\mathcal{J}\left((u*\alpha_n)(t)\right), \quad t>t_0,\ n\geq n_0 \tag{2.12}$$

(the commutativity between the bounded operator $\mathcal{J}$ and the derivative $\frac{\mathrm{d}}{\mathrm{d}t}$ is obvious).

Remember also that $\mathcal{J}$ is isometric, X onto X'' (as X is reflexive). Then $\mathcal{J}^{-1}$ is a linear continuous mapping, $X''\to X$. We can write, as follows from (2.12), the equality

$$A''\mathcal{J}(u*\alpha_n)(t) = \mathcal{J}\left(\frac{\mathrm{d}}{\mathrm{d}t}(u*\alpha_n)\right)(t), \quad t>t_0,\ n>n_0 \tag{2.13}$$

and then

$$\left(\mathcal{J}^{-1}A''\mathcal{J}\right)(u*\alpha_n)(t) = \frac{\mathrm{d}}{\mathrm{d}t}(u*\alpha_n)(t), \quad t>t_0,\ n>n_0 . \tag{2.14}$$

■

We next prove a special case of a well-known result (see for instance Theorem II, 2.14, p. 56 in [4]).

Proposition 2.1 *Let X be a reflexive B-space, T a linear closed operator with dense domain $\mathcal{D}(T)$ in X. Then, if $\mathcal{J}$ is the canonical mapping of X onto X'', we have the equality*

$$\mathcal{J}^{-1}T''\mathcal{J} = T . \tag{2.15}$$

(Note that from Corollary 1.3 of this chapter, it follows that the domain of the dual operator T' is dense in X'; this ensures the existence of the second dual operator T''.)

Proof The proof will be given in a few steps.

(a) The operator $U = \mathcal{J}^{-1}T''\mathcal{J}$ is a closed operator in X (note that $\mathcal{D}(U) = \{x \in X,\ \mathcal{J}x \in \mathcal{D}(T'')\}$.

Remember that T'' (as the dual of T') is a closed operator. Then, take a sequence (x_n) in $\mathcal{D}(U)$, where $x_n \to x_0$ in X while $Ux_n \to y_0$ in X. It follows that

$$\mathcal{J}x_n \to \mathcal{J}x_0 \text{ in } X'' \quad \text{and} \quad T''(\mathcal{J}x_n) = \mathcal{J}(Ux_n) \to \mathcal{J}y_0 \text{ in } X''$$

from the closedness of T'' we infer that: $\mathcal{J}x_0 \in \mathcal{D}(T'')$ and $T''\mathcal{J}x_0 = \mathcal{J}y_0$, that is $\mathcal{J}^{-1}T''\mathcal{J}x_0 = y_0$ or $Ux_0 = y_0$. Therefore, $x_0 \in \mathcal{D}(U)$ and $Ux_0 = y_0$, which gives closedness of U.

(b) The operator U is a closed linear extension of T.

Take in fact $x \in \mathcal{D}(T)$. Then $(\mathcal{J}x)(T'y') = (T'y')(x)$, $\forall y' \in \mathcal{D}(T')$. Also, $(T'y')(x) = y'(Tx)$ as $x \in \mathcal{D}(T)$ and $y' \in \mathcal{D}(T')$. Therefore

$$(\mathcal{J}x)(T'y') = y'(Tx) , \quad \forall x \in \mathcal{D}(T),\ \forall y' \in \mathcal{D}(T') . \tag{2.16}$$

We derive the estimate

$$|(\mathcal{J}x)(T'y')| \le \|Tx\|\,\|y'\| , \quad x \in \mathcal{D}(T),\ \forall y' \in \mathcal{D}(T') \tag{2.17}$$

which indicates that the operator $(\mathcal{J}x)\circ T'$ is linear continuous on $\mathcal{D}(T')$, and, as seen in (15.6), Chapter II, this in turn implies that $\mathcal{J}x \in \mathcal{D}(T'')$, and therefore $x \in \mathcal{D}(U)$. Thus, $\mathcal{D}(T) \subset \mathcal{D}(U)$.

Consider now, on the other hand, $x \in \mathcal{D}(U)$ and $y' \in \mathcal{D}(T')$. We have

$$y'\left(\mathcal{J}^{-1}T''\mathcal{J}x\right) = \mathcal{J}\left(\mathcal{J}^{-1}T''\mathcal{J}x\right)(y') \tag{2.18}$$

because of the formula $y'(z) = (\mathcal{J}z)(y'),\ \forall z \in X,\ y' \in X'$.

Therefore, from (2.18) one obtains, $\forall x \in \mathcal{D}(U),\ \forall y' \in \mathcal{D}(T')$

$$y'(Ux) = (T''\mathcal{J}x)(y') \ . \tag{2.19}$$

Furthermore we also have the relations, $\forall x \in \mathcal{D}(U) = \mathcal{D}(T''\mathcal{J}),\ \forall y' \in \mathcal{D}(T')$

$$(T'y')(x) = (\mathcal{J}x)(T'y') = T''(\mathcal{J}x)(y') \ . \tag{2.20}$$

Substituting in (2.19) one obtains, accordingly, the equality

$$y'(Ux) = (T'y')(x) \ , \quad \forall x \in \mathcal{D}(U),\ \forall y' \in \mathcal{D}(T') \ . \tag{2.21}$$

From (2.21) it follows, taking $x \in \mathcal{D}(T)$, which is contained in $\mathcal{D}(U)$, the equality

$$y'(Ux) = (T'y')(x) = y'(Tx) \ , \quad \forall y' \in \mathcal{D}(T') \ . \tag{2.22}$$

As $\mathcal{D}(T')$ is dense in X', this last result implies, for $x \in \mathcal{D}(T)$, that $Tx = Ux$.

Therefore, the operator U is indeed a closed linear extension of the operator T.

(c) In this last part of the proof we show that $G_U \subset G_T$ (in the space $X \times X$). Suppose *not*; then $\exists (u,v) \in G_U,\ (u,v) \notin G_T$. As G_T is closed in $X \times X$, there exists an element $z' \in (X \times X)'$ such that $z'((x,Tx)) = 0,\ \forall x \in \mathcal{D}(T)$ while $z'((u,v)) \neq 0$.

Let elements $x' \in X',\ y' \in X'$ be defined by

$$x'(x) = z'((x,\mathbf{0})) \ , \quad y'(y) = z'((\mathbf{0},y)) \ , \quad \forall x \in X,\ \forall y \in X \ . \tag{2.23}$$

It follows obviously that

$$z'((u,v)) = x'(u) + y'(v) \neq 0 \tag{2.24}$$

while

$$x'(x) + y'(Tx) = z'((x,Tx)) = 0 \ , \quad \forall x \in \mathcal{D}(T) \ . \tag{2.25}$$

Thus, $x'(x) = -y'(Tx)$, $\forall x \in \mathcal{D}(T)$, hence $y' \in \mathcal{D}(T')$ and

$$T'y' = -x' \ . \tag{2.26}$$

On the other hand, let us note that, as $(u, v) \in G_U$, one has

$$v = Uu = \mathcal{J}^{-1}T''\mathcal{J}u \ . \tag{2.27}$$

Use also (2.21) taking $x = u$: thus, using also (2.26)

$$y'(Uu) = (T'y')(u) = -x'(u) \ , \qquad \forall y' \in \mathcal{D}(T') \ . \tag{2.28}$$

Actually, $Uu = v$, hence (2.28) gives

$$y'(v) + x'(u) = 0 \ ,$$

contradicting (2.24). Thus

$$G_U \subset G_T \subset G_U \quad \text{and} \quad U = T \ ;$$

which is (2.15). ∎

3. We continue here the proof of Theorem 2.1 which was started in the previous section. The main result in §2 was (2.14) which, when combined with (2.15), gives the equality

$$\frac{\mathrm{d}}{\mathrm{d}t}(u * \alpha_n)(t) = A(u * \alpha_n)(t) \ , \quad \text{for } t > t_0,\ n > n_0 \ . \tag{3.1}$$

Take any $N > t_0$ and integrate by parts; one obtains

$$\int_{t_0}^{N} \mathrm{e}^{-\lambda t} \frac{\mathrm{d}}{\mathrm{d}t}(u * \alpha_n)\, \mathrm{d}t = \int_{t_0}^{N} \mathrm{e}^{-\lambda t} A(u * \alpha_n)(t)\, \mathrm{d}t = \mathrm{e}^{-\lambda N}(u * \alpha_n)(N)$$
$$- \mathrm{e}^{-\lambda t_0}(u * \alpha_n)(t_0) + \lambda \int_{t_0}^{N} \mathrm{e}^{-\lambda t}(u * \alpha_n)(t)\, \mathrm{d}t \tag{3.2}$$

which holds for $n > n_0$, $\lambda \in \mathbb{C}$.

Take then $\lambda \in \mathbb{C}$, $\operatorname{Re}\lambda > w$; as previously seen, we obtain absolute convergence of the (improper) integral

$$\int_{t_0}^{\infty} e^{-\lambda t}(u * \alpha_n)(t)\, dt \ ;$$

also, from (2.6), where $w < w_1 < \operatorname{Re}\lambda$, we get the estimate

$$\left\| e^{-\lambda N}(u * \alpha_n)(N) \right\| \le e^{-(\operatorname{Re}\lambda)N} C_{w_1} e^{N w_1} = C_{w_1} e^{N(w_1 - \operatorname{Re}\lambda)} \to 0 \tag{3.3}$$

as $N \to \infty$ (for $n > n_0$).

Thus, from the above remarks, one obtains:

$$\lim_{N\to\infty} \int_{t_0}^{N} e^{-\lambda t} A(u * \alpha_n)(t)\, dt = -e^{-\lambda t_0}(u * \alpha_n)(t_0) + \lambda \int_{t_0}^{\infty} e^{-\lambda t}(u * \alpha_n)(t)\, dt \ . \tag{3.4}$$

We need a certain quite simple but often useful result which is here stated as:

Proposition 3.1 *Let T, $\mathcal{D}(T) \subset X \to X$, be a linear closed operator in the Banach space X, and $x(\cdot)$, $a \le t < b \to X$, $b \le +\infty$ be a continuous function. Suppose that $x(t) \in \mathcal{D}(T)$, $\forall t \in [a,b)$, that the function $(Tx)(\cdot)$ is continuous on $[a,b)$ and that the improper Riemann integrals*

$$\int_a^b x(t)\, dt \quad \textit{and} \quad \int_a^b (Tx)(t)\, dt$$

exists. Then

$$\int_a^b x(t)\, dt \in \mathcal{D}(T) \quad \textit{and} \quad T\int_a^b x(t)\, dt = \int_a^b (Tx)(t)\, dt \ . \tag{3.5}$$

(Thus, one can loosely say that closed operators commute with Riemann integrals; we already know that linear continuous operators commute with R-integrals (see (3.8) in Chapter V).)

Proof We only consider here the case $b = +\infty$, $a = t_0$. Take $N > t_0$ and a partition π of the interval $[t_0, N]$; we have, for corresponding Riemann sums

$$f_n = \sum_{i=0}^{n-1} x(\tau_i)(t_{i+1} - t_i) \in \mathcal{D}(T) \quad \text{and} \quad g_n = \sum_{i=0}^{n-1} (Tx)(\tau_i)(t_{i+1} - t_i) = Tf_n .$$

As $\|\pi\| \to 0$ and $n \to \infty$, one obtains

$$\lim_{n\to\infty} f_n = \int_{t_0}^{N} x(t)\,\mathrm{d}t \quad \text{and} \quad \lim_{n\to\infty} Tf_n = \int_{t_0}^{N} (Tx)(t)\,\mathrm{d}t .$$

As T is a closed operator we obtain

$$\int_{t_0}^{N} x(t)\,\mathrm{d}t \in \mathcal{D}(T) \quad \text{and} \quad T\int_{t_0}^{N} x(t)\,\mathrm{d}t = \int_{t_0}^{N} (Tx)(t)\,\mathrm{d}t .$$

Next we shall handle the case of the improper integral $\int_{t_0}^{\infty} x(t)\,\mathrm{d}t$. We have the sequence of elements $x_N = \int_{t_0}^{N} x(t)\,\mathrm{d}t$, where $x_N \in \mathcal{D}(T)$, $N > t_0$ is a natural number; also

$$Tx_N = \int_{t_0}^{N} (Tx)(t)\,\mathrm{d}t .$$

By assumption,

$$\lim_{N\to\infty} x_N = \int_{t_0}^{\infty} x(t)\,\mathrm{d}t$$

while

$$\lim_{N\to\infty} Tx_N = \int_{t_0}^{\infty} (Tx)(t)\,\mathrm{d}t .$$

Again, T is a closed operator, so that we obtain

$$\int_{t_0}^{\infty} x(t)\,\mathrm{d}t \in \mathcal{D}(T) \quad \text{and} \quad T\int_{t_0}^{\infty} x(t)\,\mathrm{d}t = \int_{t_0}^{\infty} (Tx)(t)\,\mathrm{d}t$$

which is (3.5) in our special case.

We shall apply this result in (3.4), taking $x(t) = \mathrm{e}^{-\lambda t}(u * \alpha_n)(t)$, $t \geq t_0$, and $T = A$; here $(Ax)(t) = \mathrm{e}^{-\lambda t} A(u * \alpha_n)(t) = \mathrm{e}^{-\lambda t} \dfrac{\mathrm{d}}{\mathrm{d}t}(u * \alpha_n)(t)$ is a continuous function on $[t_0, \infty)$.

The improper integral $\int_{t_0}^{\infty} \mathrm{e}^{-\lambda t}(u * \alpha_n)(t)\,\mathrm{d}t$ (for $\operatorname{Re}\lambda > w$ which is here assumed), exists.

Also, as seen in (3.4), the improper integral

$$\int_{t_0}^{\infty} A\left(\mathrm{e}^{-\lambda t}(u * \alpha_n)(t)\right)\mathrm{d}t$$

also exists, and equals the right-hand side of (3.4).

Therefore, from Proposition 3.1, one derives:

$$\int_{t_0}^{\infty} \mathrm{e}^{-\lambda t}(u * \alpha_n)(t)\,\mathrm{d}t \in \mathcal{D}(A)$$

and also

$$A\int_{t_0}^{\infty} \mathrm{e}^{-\lambda t}(u * \alpha_n)(t)\,\mathrm{d}t = -\mathrm{e}^{-\lambda t_0}(u * \alpha_n)(t_0) + \lambda \int_{t_0}^{\infty} \mathrm{e}^{-\lambda t}(u * \alpha_n)(t)\,\mathrm{d}t \tag{3.6}$$

where $\operatorname{Re}\lambda > w$, $n > n_0$.

Actually, we can write (3.6) in the form

$$(\lambda - A)\int_{t_0}^{\infty} \mathrm{e}^{-\lambda t}(u * \alpha_n)(t)\,\mathrm{d}t = \mathrm{e}^{-\lambda t_0}(u * \alpha_n)(t_0)\,, \quad n > n_0,\ \operatorname{Re}\lambda > w\,. \tag{3.7}$$

We next have:

Proposition 3.2 *The following holds:*

$$\lim_{n\to\infty}(u*\alpha_n)(t_0)=u(t_0)\,,\qquad \lim_{n\to\infty}\int_{t_0}^{\infty}\mathrm{e}^{-\lambda t}(u*\alpha_n)(t)\,\mathrm{d}t=\int_{t_0}^{\infty}\mathrm{e}^{-\lambda t}u(t)\,\mathrm{d}t \tag{3.8}$$

(for $\operatorname{Re}\lambda>w$*).*

Proof If we use, for instance, definition (2.4) for the convolution $(u*\alpha_n)(\cdot)$ we obtain

$$(u*\alpha_n)(t_0)=\int_{t_0-\frac{1}{n}}^{t_0+\frac{1}{n}}u(s)\alpha_n(t_0-s)\,\mathrm{d}s\,,$$

where $n>n_0,\ t_0>\dfrac{1}{n_0}$.

Hence we get,

$$(u*\alpha_n)(t_0)-u(t_0)=\int_{t_0-\frac{1}{n}}^{t_0+\frac{1}{n}}\left\{u(s)\alpha_n(t_0-s)-u(t_0)\alpha_n(t_0-s)\right\}\mathrm{d}s$$

$$=\int_{t_0-\frac{1}{n}}^{t_0+\frac{1}{n}}\left[u(s)-u(t_0)\right]\alpha_n(t_0-s)\,\mathrm{d}s$$

and therefore the estimate

$$\left\|(u*\alpha_n)(t_0)-u(t_0)\right\|\le\sup_{|t_0-s|\le\frac{1}{n}}\left\|u(s)-u(t_0)\right\|\to 0\quad\text{as}\quad n\to\infty\,.$$

Next, in order to prove the second limit in (3.8) we shall use the so-called 'dominated convergence theorem'. This is a classical result for Lebesgue integrals; it is also true for the corresponding concept in vector-valued integration, the so-called 'Bochner integral' (see for instance [5], Theorem 3.7.9, p. 83), and the improper Riemann integral which appears in (3.8) is a special case of Bochner integral (the proof is not given here, for quite obvious reasons!!).

Therefore, we first note that

$$\lim_{n\to\infty} \mathrm{e}^{-\lambda t}(u*\alpha_n)(t) = \mathrm{e}^{-\lambda t}u(t)\,, \qquad \forall t > t_0$$

(the proof given above for $t = t_0$ works equally well for all $t > t_0$).

Afterwards, we need an estimate for $\left\|\mathrm{e}^{-\lambda t}(u*\alpha_n)(t)\right\|$, when $\operatorname{Re}\lambda > w$. Take w_1 such that $w < w_1 < \operatorname{Re}\lambda$ and use (2.6); we obtain $\left\|(u*\alpha_n)(t)\right\| \le C_{w_1}\mathrm{e}^{tw_1}$ for large t, therefore $\left\|\mathrm{e}^{-\lambda t}(u*\alpha_n)(t)\right\| \le C_{w_1}\exp\bigl(t(w_1 - \operatorname{Re}\lambda)\bigr)$ (see (2.7) above) for $n \ge n_0$, and

$$\int_{t_0}^{\infty} \mathrm{e}^{t(w_1 - \operatorname{Re}\lambda)}\,\mathrm{d}t < \infty\,.$$

We can say therefore that Proposition 3.2 has been proved. ■

We are now ready for:

Proposition 3.3 *The integral* $\int_{t_0}^{\infty} \mathrm{e}^{-\lambda t}u(t)\,\mathrm{d}t$ *for* $\operatorname{Re}\lambda > w$ *belongs to the domain* $\mathcal{D}(A)$ *and the formula*

$$(\lambda I - A)\int_{t_0}^{\infty} \mathrm{e}^{-\lambda t}u(t)\,\mathrm{d}t = \mathrm{e}^{-\lambda t_0}u(t_0) \tag{3.9}$$

holds true.

Proof We use formula (3.7) and Proposition 3.2. The operator $(\lambda I - A)$ is closed as is readily seen. Then (3.9) follows immediately.

Then we have:

Proposition 3.4 *The integral:* $\int_{0}^{\infty} \mathrm{e}^{-\lambda t}u(t)\,\mathrm{d}t$ *for* $\operatorname{Re}\lambda > w$ *belongs to the domain* $\mathcal{D}(A)$ *and the formula*

$$(\lambda I - A)\int_0^\infty e^{-\lambda t}u(t)\,dt = \mathbf{0} \tag{3.10}$$

holds true.

In fact, in (3.9) we consider $\lim_{t_0 \to 0}$, using again the closedness of the operator $(\lambda I - A)$ as well as the assumption $u(0) = \mathbf{0}$ (in (2.3)). ■

Corollary *If* $F(\lambda) = \int_0^\infty e^{-\lambda t}u(t)\,dt$, $\operatorname{Re}\lambda > w \to X$, *is* $\neq \mathbf{0}$ *for some* λ, *then this* λ *is an eigenvalue of the operator A.*

From the assumption of Theorem 2.1 it then follows that the set, $\{\lambda,\ \operatorname{Re}\lambda > w,\ F(\lambda) \neq \mathbf{0}\}$ *is not dense in* $\operatorname{Re}\lambda > w$.

Therefore, the closure $\overline{\{\lambda,\ \operatorname{Re}\lambda > w,\ F(\lambda) \neq \mathbf{0}\}}$ is not the whole half-plane $\{\lambda,\ \operatorname{Re}\lambda > w\}$ and the set $\{\lambda,\ \operatorname{Re}\lambda > w,\ F(\lambda) = \mathbf{0}\}$ contains an open disc in $\mathbb{C}$.

On the other hand, the function $F(\lambda)$ is the Laplace transform of the X-valued function $u(t)$ and as such it is an analytic function of λ (see again [5]). As $F(\lambda) = \mathbf{0}$ on a set $\{\lambda \in \mathbb{C},\ |\lambda - \lambda_0| < r\}$, it follows, as is well-known, that $F(\lambda) = 0\ \forall\lambda,\ \operatorname{Re}\lambda > w$.

By another well-known property of the Laplace transform, we get that $u(t)$, the functions whose Laplace transform is $F(\lambda)$, is itself $= \mathbf{0}$ for all $t \geq 0$.

The proof of Theorem 2.1 is thus terminated. ■

Chapter XI

Uniqueness of bounded ultraweak solutions in Hilbert space

Introduction

In this chapter we present an extension to ultraweak solutions of a quite simple result concerning uniqueness of (regular) solutions, bounded over the real line, of the equation $u'(t) = Au(t)\ t \in \mathbb{R}$, where A is a symmetric operator in Hilbert space (see [17], Theorem 1.3, p. 74).

1. The underlying Hilbert space is denoted by H. We consider a linear operator A, $\mathcal{D}(A) \subset H \to H$; we assume $\mathcal{D}(A)$ is dense in H and $A \subset A^*$, thus A is a symmetric operator in H (see Proposition 5.2, Chapter II). As A^* is closed operator, it follows that A is closable; its closure $\overline{A}$ is then equal to A^{**} (Corollary to Theorem 11.1, Chapter II). As $A \subset A^*$ it follows that $\overline{A} \subset A^*$ too. By Proposition 11.1, Chapter II, we also have that $(\overline{A})^* = A^*$ and therefore we have $\overline{A} \subset A^* = (\overline{A})^*$, which means that the closure $\overline{A}$ is again a symmetric operator in H.

We next define ultraweak solutions of the differential equation $u'(t) = Au(t)$, $t \in \mathbb{R}$ in the space H (see for instance [21-I], Chapter X).

First a class of vector-valued test-functions, here denoted by $K_{A^*}(\mathbb{R})$, is composed of functions $\varphi(\cdot)$, $\mathbb{R} \to \mathcal{D}(A^*)$, such that $\varphi(\cdot) \in C^1(\mathbb{R}; H)$, $(A^*\varphi)(\cdot) \in C(\mathbb{R}; H)$, $\varphi(t) = \mathbf{0}$ for $|t| \geq \bar{t}_\varphi$.

Next, consider functions $u(\cdot) \in C(\mathbb{R}; H)$, such that

$$\int_{\mathbb{R}} (u(t),\ \varphi'(t) + (A^*\varphi)(t))_H \,\mathrm{d}t = 0\,, \qquad \forall \varphi(\cdot) \in K_{A^*}(\mathbb{R}) \tag{1.1}$$

holds true.

(These are the 'ultraweak' solutions in the present chapter.)

Our aim here is to establish an ultraweak version of Theorem 1.3, p. 74 in [17]:

Theorem 1.1 *Let us assume (1.1) and also the conditions:*

(i) $\mathbf{0} = h \in \mathcal{D}(\overline{A})$ *iff* $\overline{A}h = \mathbf{0}$; (1.2)

(ii) $\sup_{t\in\mathbb{R}} \int_t^{t+1} \|u(\sigma)\|^2 \, \mathrm{d}\sigma < \infty$ *(the S^2 boundedness-condition).* (1.3)

Then $u(t) = \mathbf{0}$, $\forall t \in \mathbb{R}$.

The proof is based on the simple idea of reducing ultraweak solutions to regular, ordinary, solutions by the mollification procedure, and then applying the previously mentioned Theorem 1.3 in [17].

Thus, let us consider a sequence $\{\alpha_n(\cdot)\}$ of scalar-valued non-negative C^1-functions, vanishing for $|t| > \frac{1}{n}$, such that $\int_{-\infty}^{\infty} \alpha_n(\sigma)\,\mathrm{d}\sigma = 1$, $\forall n \in \mathbb{N}$. Then define convolutions $(u * \alpha_n)(\cdot)$ given by the usual formula

$$(u*\alpha_n)(t) = \int_{\mathbb{R}} u(s)\alpha_n(t-s)\,\mathrm{d}s = \int_{t-\frac{1}{n}}^{t+\frac{1}{n}} u(s)\alpha_n(t-s)\,\mathrm{d}s\,, \qquad t \in \mathbb{R}\,. \tag{1.4}$$

Our first remark is given as:

Lemma 1.1 *All functions $(u*\alpha_n)(\cdot)$ satisfy (1.3).*

Proof We have in fact the following simple estimates:

$$\left(\int_t^{t+1} \|(u*\alpha_n)(\sigma)\|^2\,\mathrm{d}\sigma\right)^{\frac{1}{2}} = \left(\int_t^{t+1} \left\|\int_{\mathbb{R}} u(s)\alpha_n(\sigma-s)\,\mathrm{d}s\right\|^2 \mathrm{d}\sigma\right)^{\frac{1}{2}}$$

$$\leq \left(\int_t^{t+1}\left(\int_{\mathbb{R}} \|u(s)\|\alpha_n(\sigma-s)\,\mathrm{d}s\right)^2 \mathrm{d}\sigma\right)^{\frac{1}{2}} = \left(\int_t^{t+1}\left(\int_{\mathbb{R}} \|u(\sigma-\tau)\|\alpha_n(\tau)\,\mathrm{d}\tau\right)^2 \mathrm{d}\sigma\right)^{\frac{1}{2}}. \tag{1.5}$$

If we consider the (scalar-valued) function $f(\sigma,\tau)=\|u(\sigma-\tau)\|\alpha_n(\tau)\chi_t(\sigma)$, $\chi_t(\sigma)=1$ for $t\le\sigma\le t+1$, $=0$ elsewhere, where $\sigma\in\mathbb{R}$, $\tau\in\mathbb{R}$, we get the right-hand side in (1.5) as

$$\left(\int_{\mathbb{R}}\left(\int_{\mathbb{R}} f(\sigma,\tau)\,\mathrm{d}\tau\right)^2 \mathrm{d}\sigma\right)^{\frac{1}{2}}$$

which is estimated (from before) in view of Minkowski's inequality (see [9], p. 4) by

$$\int_{\mathbb{R}}\left(\int_{\mathbb{R}} f^2(\sigma,\tau)\,\mathrm{d}\sigma\right)^{\frac{1}{2}} \mathrm{d}\tau$$

which in turn equals

$$\int_{\mathbb{R}}\left(\int_{\mathbb{R}} \|u(\sigma-\tau)\|^2\alpha_n^2(\tau)\chi_t^2(\sigma)\,\mathrm{d}\sigma\right)^{\frac{1}{2}} \mathrm{d}\tau=\int_{\mathbb{R}}\left(\int_t^{t+1}\|u(\sigma-\tau)\|^2\,\mathrm{d}\sigma\right)\cdot\alpha_n(\tau)\,\mathrm{d}\tau\,. \tag{1.6}$$

Next,

$$\int_t^{t+1}\|u(\sigma-\tau)\|^2\,\mathrm{d}\sigma=\int_{t-\tau}^{t-\tau+1}\|u(s)\|^2\,\mathrm{d}s\le L=\sup_{t\in\mathbb{R}}\int_t^{t+1}\|u(\sigma)\|^2\,\mathrm{d}\sigma \tag{1.7}$$

(see (1.3)).

Therefore, we obtain, from (1.5)-(1.7), that

$$\left(\int_t^{t+1}\|(u*\alpha_n)(\sigma)\|^2\,\mathrm{d}\sigma\right)^{\frac{1}{2}}\le L\int_{\mathbb{R}}\alpha_n(\tau)\,\mathrm{d}\tau=L\,,\qquad \forall t\in\mathbb{R} \tag{1.8}$$

which proves the lemma. ■

Now we show that functions $u*\alpha_n$ are actually regular solutions to the equation

$$(u*\alpha_n)'(t)=\overline{A}(u*\alpha_n)(t)\,,\qquad t\in\mathbb{R}\,. \tag{1.9}$$

In fact, it is obvious from (1.4) that the derivative $(u*\alpha_n)'(t)$ will exist ($\forall t\in\mathbb{R}$), being equal to $\int_{\mathbb{R}} u(s)\alpha_n'(t-s)\,\mathrm{d}s$.

Next, we use (1.1), taking as test-functions, $\forall t\in\mathbb{R}$, $\varphi_{n,t}(s)=\alpha_n(t-s)h$, where $h\in\mathcal{D}(A^*)$.

(We check readily that $\varphi_{n,t}(\cdot)\in K_{A^*}(\mathbb{R})$, $\forall t\in\mathbb{R}$, $\forall n\in\mathbb{N}$; for instance $\varphi_{n,t}(s)\in\mathcal{D}(A^*)$, $\forall s\in\mathbb{R}$, as $h\in\mathcal{D}(A^*)$; $\varphi_{n,t}(s)\in C^1(\mathbb{R},H)$ as $s\to\alpha_n(t-s)$ belongs to $C^1(\mathbb{R})$; $A^*\varphi_{n,t}(\cdot)=\alpha_n(t-s)A^*h\in C(\mathbb{R};H)$; $\varphi_{n,t}(s)=\mathbf{0}$ for $|t-s|>\frac{1}{n}$, that is for $s\notin\left[t-\frac{1}{n},\ t+\frac{1}{n}\right]$.)

With these remarks, substituting in (1.1), we obtain the following

$$\int_{\mathbb{R}}\left(u(s),\ \varphi_{n,t}'(s)+A^*\varphi_{n,t}(s)\right)_H\mathrm{d}s=0\,,\qquad \forall n\in\mathbb{N},\ \forall t\in\mathbb{R} \tag{1.10}$$

and consequently

$$\int_{\mathbb{R}}\left(u(s),\ -\alpha_n'(t-s)h+\alpha_n(t-s)A^*h\right)_H\mathrm{d}s=0\,,\qquad \forall h\in\mathcal{D}(A^*),\ \forall t\in\mathbb{R} \tag{1.11}$$

which is also written as follows:

$$\left(\int_{\mathbb{R}}\alpha_n(t-s)u(s)\,\mathrm{d}s,\ A^*h\right)_H=\left(\int_{\mathbb{R}}\alpha_n'(t-s)u(s)\,\mathrm{d}s,\ h\right),\qquad \forall h\in\mathcal{D}(A^*),\ \forall t\in\mathbb{R} \tag{1.12}$$

using obvious properties of the R-integral in relation with the H-inner product.

If now one uses the *definition* on the second adjoint operator A^{**} we get from (1.12):

$$\int_{\mathbb{R}}\alpha_n(t-s)u(s)\,\mathrm{d}s\quad\text{belongs to }\ \mathcal{D}(A^{**})$$

and

$$A^{**}\int_{\mathbb{R}}\alpha_n(t-s)u(s)\,\mathrm{d}s=\int_{\mathbb{R}}\alpha_n'(t-s)u(s)\,\mathrm{d}s \tag{1.13}$$

holds true.

As seen previously, the second adjoint A^{**} equals the closure $\overline{A}$ and we have also seen above that

$$\int_{\mathbb{R}} \alpha_n'(t-s)u(s)\,\mathrm{d}s = (u*\alpha_n)'(t)$$

whence the equality

$$(u*\alpha_n)'(t) = \overline{A}(u*\alpha_n)(t)\,, \quad \forall t \in \mathbb{R},\ \forall n \in \mathbb{N}\,. \tag{1.14}$$

We are thus entitled to use assumption (i), stating that $\lambda = 0$ is not an eigenvalue of the operator $\overline{A}$; estimate (ii) (1.3) for all $u*\alpha_n$, as established in Lemma 1.1, the symmetry property of operator $\overline{A}$, as noted at the beginning, and Theorem 1.3 p. 74 of [17], to get

$$(u*\alpha_n)(t) = \mathbf{0}\,, \quad \forall t \in \mathbb{R},\ \forall n \in \mathbb{N}\,.$$

Finally, it is easy to see that from the continuity of $u(\cdot)$

$$(u*\alpha_n)(t) \to u(t) \quad \text{as } n \to \infty,\ \forall t \in \mathbb{R}$$

so that $u(t) = \mathbf{0}$, $\forall t \in \mathbb{R}$ follows. ∎

Note One should compare the previous remark with the paper [15], where a similar discussion takes place, essentially in a more special situation.

Chapter XII

Almost-periodic solutions I

Introduction

In this chapter we shall explain a simple result concerning the almost-periodicity of solutions for certain 'abstract' differential equations, that is, here, differential equations in Banach space with unbounded operators as coefficients. More precisely, we shall see that, under appropriate conditions, weakly almost-periodic solutions are also strongly almost-periodic. Preliminary general considerations on differential equations are also included.

1. Consider an arbitrary Banach space X; then consider also a one-parameter family of linear continuous mappings of X into itself, the parameter t runs over the real line:

$$t \in \mathbb{R} \to G(t) \in \mathcal{L}(X) \,. \tag{1.1}$$

We assume that the family $G(t)$ is a group of operators:

$$G(t)G(s) = G(s)G(t) = G(t+s)\,, \qquad \forall s,t \in \mathbb{R};\ G(0) = I\,(\text{dentity}) \tag{1.2}$$

holds true.

Thus, each operator $G(t)$ has an inverse operator which is $G(-t)$.

We also assume 'strong continuity' of the group $G(t)$: the function

$$t \in \mathbb{R} \to G(t)x \in X \tag{1.3}$$

is continuous $\forall x \in X$.

Note It follows from (1.3) that $\lim_{t\to 0} G(t)x = G(0)x = x$, $\forall x \in X$. Conversely, if we assume this property we can derive (1.3) as a corollary:

$$\lim_{h\to 0} G(t+h)x = \lim_{h\to 0} G(h)G(t)x = G(t)x\,, \qquad \forall x \in X,\ \forall t \in \mathbb{R}\,.$$

Note also the following:

Proposition 1.1 *If the group of operators $G(t)$ satisfies (1.1)-(1.3), it follows that*

$$\sup_{t\in[a,b]} \|G(t)\| < +\infty \tag{1.4}$$

for any compact subinterval $[a,b]$ of $\mathbb{R}$.

Proof If $x \in X$, we see that, from continuity of the function $t \in \mathbb{R} \to G(t)x$, one can infer

$$\sup_{a\le t\le b} \|G(x)x\| < +\infty \ .$$

Then one applies the uniform boundedness theorem (Chapter II-2).

Next, as in Chapter V – on C_0-(semi)groups of operators – we can define the fundamental concept of an 'infinitesimal generator' for a group $G(t)$.

We consider the derivative for $t = 0$:

$$\lim_{h\to 0} \frac{1}{h}[G(h)x - x] , \quad \text{where } \ x \in X \ . \tag{1.5}$$

In general, this derivative will exist for a (strict) subset of X. Then one defines a set $\mathcal{D}$ by

$$\mathcal{D} = \left\{ x \in X,\ \frac{\mathrm{d}}{\mathrm{d}t} G(t)x\Big|_{t=0} \ \text{exists} \right\} \tag{1.6}$$

and, accordingly, an operator A, $\mathcal{D} \to X$, is defined by

$$Ax = \lim_{\eta\to 0} \frac{1}{\eta}[G(\eta)x - x] , \quad \forall x \in \mathcal{D} \ . \tag{1.7}$$

It is readily seen that $\mathcal{D}$ is a linear subset of X and that A is a linear operator, $\mathcal{D}(A) = \mathcal{D}$.

This is in fact the so-called 'infinitesimal operator' of $G(t)$.

It follows that, for $x \in \mathcal{D}$, the function

$$t \in \mathbb{R} \to G(t)x \in X \tag{1.8}$$

is a solution to an 'abstract differential equation;

$$\frac{d}{dt}G(t)x = AG(t)x = G(t)Ax\,, \qquad \forall t \in \mathbb{R},\ \forall x \in \mathcal{D}(A) \tag{1.9}$$

(we can thus say that $\mathcal{D}(A)$ is invariant under $G(t)$; the operators $G(t)$ and A commute on $\mathcal{D}(A)$).

It follows also that the derivative $\frac{d}{dt}G(t)x$ (for $x \in \mathcal{D}$) is a continuous function, $\mathbb{R} \to X$, and the function $u(t) = G(t)x$, $x \in \mathcal{D}$, $t \in \mathbb{R}$, is a solution to the abstract Cauchy problem

$$u'(t) = Au(t)\,, \qquad u(0) = x,\ \forall t \in \mathbb{R}. \tag{1.10}$$

In order to establish (1.9) we consider the differential quotient $(\forall t \in \mathbb{R})$ $\frac{1}{\eta}[G(t+\eta)x - G(t)x]$ which equals $G(t)\frac{1}{\eta}[G(\eta)x - x]$ and also $\frac{G(\eta)-I}{\eta}G(t)x$.

If $x \in \mathcal{D}$, $\frac{1}{\eta}[G(\eta)x - x] \to Ax$, hence $G(t)\frac{1}{\eta}[G(\eta)x - x] \to G(t)Ax$. This means also that $\lim\limits_{\eta\to 0}\frac{G(\eta)-I}{\eta}G(t)x$ exists and accordingly (for $x \in \mathcal{D}$), $G(t)x \in \mathcal{D}$ too and the equality

$$G(t)Ax = AG(t)x \tag{1.11}$$

holds true, as well as (1.9)-(1.10). Also, as $\frac{d}{dt}G(t)x = G(t)Ax$ (for $x \in \mathcal{D}$), we infer that the derivative $u'(t)$ is continuous, $\mathbb{R} \to X$. Consider next the C_0-semigroup $T(t)$ which is the restriction of the group $G(t)$ to the half-line $[0,\infty)$.

Let us denote by A^+ the infinitesimal generator of $T(t)$. We have:

Proposition 1.2 *The operators A and A^+ coincide.*

Proof Let $x \in \mathcal{D} = \mathcal{D}(A)$. Then: $\lim\limits_{\eta\to 0}\frac{G(\eta)x - x}{\eta}$ exists $(= Ax)$. Hence,

$$\lim_{\substack{\eta\to 0\\ \eta>0}}\frac{G(\eta)x - x}{\eta} = \lim_{\substack{\eta\to 0\\ \eta>0}}\frac{T(\eta)x - x}{\eta}$$

exists too, and equals A^+x. Thus, on $\mathcal{D}$, $A^+ = A$ and we can say that $A \subset A^+$.

Take now $x \in \mathcal{D}(A^+)$, so that $\lim\limits_{\substack{\eta\to 0\\ \eta>0}} \dfrac{G(\eta)x-x}{\eta}$ exists $(= A^+x)$. Then

$$\lim_{\substack{\eta\to 0\\ \eta>0}} G(-\eta)\frac{G(\eta)x-x}{\eta}$$

exists and equals

$$A^+x = \lim_{\substack{\eta\to 0\\ \eta>0}} \frac{x-G(-\eta)x}{\eta}.$$

In fact, if we denote $z_\eta = \dfrac{1}{\eta}[G(\eta)x-x]$ we have $z_\eta \to A^+x$ as $\eta \to 0,\ \eta > 0$; then

$$G(-\eta)z_\eta - A^+x = G(-\eta)\left[z_\eta - A^+x\right] + G(-\eta)A^+x - A^+x\ . \tag{1.12}$$

Estimating in (1.12) one obtains (for $|\eta| \le 1$)

$$\begin{aligned}
&\left\|G(-\eta)z_\eta - A^+x\right\| \\
&\quad\le \|G(-\eta)\|\left\|z_\eta - A^+x\right\| + \left\|G(-\eta)A^+x - A^+x\right\| \\
&\quad\le \sup_{|\eta|\le 1}\|G(-\eta)\|\left\|z_\eta - A^+x\right\| + \left\|G(-\eta)A^+x - A^+x\right\| \to 0\ , \qquad \text{as } \eta\to 0,\ \eta>0.
\end{aligned} \tag{1.13}$$

On the other hand,

$$G(-\eta)\frac{G(\eta)x-x}{\eta} = \frac{G(-\eta)x-x}{-\eta}.$$

We have proved therefore that

$$\lim_{\substack{\eta\to 0\\ \eta<0}} \frac{G(\eta)x-x}{\eta}$$

exists (for $x \in \mathcal{D}(A^+)$) and equals A^+x.

As both the right and the left derivative at $t = 0$ of $G(t)x$ exist and are equal (to A^+x), we infer that

$$\frac{\mathrm{d}}{\mathrm{d}t}G(t)x\Big|_{t=0} \quad \text{exists and equals } A^+x \quad (\text{for } x \in \mathcal{D}(A^+))$$

which means that $x \in D(A)$ and $Ax = A^+x$; accordingly $A^+ \subset A$.

Thus, $A = A^+$ which is Proposition 1.2. ■

Proposition 1.2 has two important corollaries, namely:

The infinitesimal generator A of the group $G(t)$ is a closed operator with dense domain.

This follows from the above proposition and Proposition 3.3-3.4 in Chapter V.

We shall close this section with a representation result for solutions of the equation

$$v'(t) = Av(t) + g(t)\,, \qquad t \in \mathbb{R} \tag{1.14}$$

where A is again the infinitesimal generator of the group $G(t)$. This will be essential when we explain the almost-periodicity result in the next section.

Thus, let us state (and then prove):

Proposition 1.3 *Let A be the generator of the group $G(t)$; $g(\cdot)$, $t \in \mathbb{R} \to X$ be a continuous function; $v(\cdot)$, $\mathbb{R} \to \mathcal{D}(A)$ be a $C^1(\mathbb{R}\,; X)$-function, solution of the differential equation*

$$v'(t) = Av(t) + g(t)\,, \qquad t \in \mathbb{R}\,. \tag{1.15}$$

Then $v(\cdot)$ is represented by the formula

$$v(t) = G(t)v(0) + \int_0^t G(t-\sigma)g(\sigma)\,\mathrm{d}\sigma\,, \qquad \forall t \in \mathbb{R}\,. \tag{1.16}$$

Remark This result appears in [21-II - Lemma 1, p. 37].

Proof Let us write (1.15) as

$$v'(\sigma) = Av(\sigma) + g(\sigma) , \qquad \sigma \in \mathbb{R}$$

and then apply here, given $t \in \mathbb{R}$, the operator $G(t-\sigma)$. One obtains

$$G(t-\sigma)v'(\sigma) - G(t-\sigma)(Av)(\sigma) = G(t-\sigma)g(\sigma)$$

and after integration (from 0 to t), we get

$$\int_0^t G(t-\sigma)v'(\sigma)\,\mathrm{d}\sigma - \int_0^t G(t-\sigma)(Av)(\sigma)\,\mathrm{d}\sigma = \int_0^t G(t-\sigma)g(\sigma)\,\mathrm{d}\sigma . \tag{1.17}$$

(Note that the functions $v'(\sigma)$, $g(\sigma)$ and $(Av)(\sigma) = v'(\sigma) - g(\sigma)$ are continuous, $\mathbb{R} \to X$; if $h(\sigma) \in C(\mathbb{R};X)$ then $\sigma \to G(-\sigma)h(\sigma) \in C(\mathbb{R};X)$ too - this is not difficult to prove; thus, the integrals in (1.17) are abstract Riemann integrals.)

Next, we note the equality

$$\frac{\mathrm{d}}{\mathrm{d}\sigma} G(t-\sigma)v(\sigma) = -AG(t-\sigma)v(\sigma) + G(t-\sigma)v'(\sigma) . \tag{1.18}$$

This is seen as follows

$$\frac{\mathrm{d}}{\mathrm{d}\sigma} G(t)G(-\sigma)v(\sigma) = G(t)\frac{\mathrm{d}}{\mathrm{d}\sigma} G(-\sigma)v(\sigma) , \qquad \forall t \in \mathbb{R};$$

$$\frac{\mathrm{d}}{\mathrm{d}\sigma} G(-\sigma)v(\sigma) = \lim_{\eta\to 0} \frac{1}{\eta}[G(-\sigma-\eta)v(\sigma+\eta) - G(-\sigma)v(\sigma)]$$

$$= \lim_{\eta\to 0} \frac{1}{\eta}[G(-\sigma-\eta)v(\sigma) - G(-\sigma)v(\sigma)]$$

$$+ \lim_{\eta\to 0} \frac{1}{\eta}\{G(-\sigma-\eta)[v(\sigma+\eta) - v(\sigma)]\} .$$

The first expression on the right-hand side is obviously

$$G(-\sigma) \lim_{\eta\to 0} - \frac{G(-\eta)v(\sigma) - v(\sigma)}{-\eta} = G(-\sigma) \quad (-Av(\sigma)) .$$

The second expression on the right-hand side is

$$\lim_{\eta\to 0} G(-\sigma)G(-\eta)\frac{v(\sigma+\eta)-v(\sigma)}{\eta} = G(-\sigma)v'(\sigma)\ ;$$

this is because of the following computation:

$$G(-\eta)\frac{1}{\eta}[v(\sigma+\eta)-v(\sigma)]-v'(\sigma)= G(-\eta)\frac{1}{\eta}[v(\sigma+\eta)-v(\sigma)]-v'(\sigma)$$
$$+G(-\eta)v'(\sigma)-v'(\sigma)\ ,$$

whence the estimate (for $|\eta|\le 1$):

$$\left\|G(-\eta)\frac{1}{\eta}[v(\sigma+\eta)-v(\sigma)]-v'(\sigma)\right\|\le \sup_{|\eta|\le 1}\|G(-\eta)\|\left\|\frac{1}{\eta}[v(\sigma+\eta)-v(\sigma)]-v'(\sigma)\right\|$$
$$+\|G(-\eta)v'(\sigma)-v'(\sigma)\| \quad \text{which} \to 0 \ \text{ as } \ \eta\to 0 \quad (\forall\sigma\in\mathbb{R}).$$

Altogether we obtain

$$\frac{\mathrm{d}}{\mathrm{d}\sigma}G(t-\sigma)v(\sigma)=-G(t-\sigma)Av(\sigma)+G(t-\sigma)v'(\sigma)$$

which is (1.18) in view of (1.11).

If one integrates (1.18) between 0 and t, one obtains – also using (1.17) –

$$v(t)-G(t)v(0)=-\int_0^t G(t-\sigma)Av(\sigma)\,\mathrm{d}\sigma+\int_0^t G(t-\sigma)v'(\sigma)\,\mathrm{d}\sigma=\int_0^t G(t-\sigma)g(\sigma)\,\mathrm{d}\sigma$$

which is (1.16). ■

2. We start this section with a few definitions and results from the theory of abstract almost periodic functions (see [8, 18]).

A continuous function $f(\cdot)$, $\mathbb{R}\to X$ (a Banach space) is called almost-periodic if, $\forall\varepsilon>0$, there exists a relatively dense set of ε-translation numbers for it - equivalently, if $\forall\varepsilon>0$, $\exists l(\varepsilon)>0$, such that in any real interval of length l there is at least one number τ, with the property that

$$\|f(t+\tau)-f(t)\|<\varepsilon, \quad \forall t\in\mathbb{R} .$$

A main assumption in this chapter is:

The operator group $G(t)$ is strongly almost-periodic, that is, $\forall x\in X$, the function $t\in\mathbb{R}\to G(t)x\in X$ is almost periodic.

We need also the concept of weakly almost-periodic functions (see [18], Chapter 11).

A function $f(\cdot)$, $\mathbb{R}\to X$ (a Banach space) is called weakly almost-periodic if, for any $x'\in X'$ (the dual space of X), the numerical-valued function

$$t\in\mathbb{R}\to x'(f(t))$$

is almost-periodic.

Thus, weakly almost-periodic functions are weakly continuous ($x'(f(t))$ is a continuous function, $\forall x'\in X'$).

Our main result in the present chapter is stated as:

Theorem 2.1 *Let X be a reflexive B-space; $G(t)$ a strongly almost-periodic group of operators (satisfying assumption (1.1)-(1.3)) and let A be its infinitesimal generator; B a linear compact operator, $X\to X$; and let $v(\cdot)$, $\mathbb{R}\to\mathcal{D}(A)$ be a solution to the differential equation*

$$v'(t)=Av(t)+Bv(t)+g(t) \tag{2.1}$$

where $g(\cdot)$ is almost-periodic, $\mathbb{R}\to X$, while $v(\cdot)$ is weakly almost-periodic, $\mathbb{R}\to X$.

Then $v(\cdot)$ is also (strongly) almost-periodic, $\mathbb{R}\to X$.

Note We assume $v(\cdot)$ to be weakly almost-periodic, but as a solution to (2.1) it is in $C^1(\mathbb{R};X)$!

In order to prove this result, we consider (2.1) as a differential equation with $Bv(\cdot)+g(\cdot)$ as the right-hand side. Therefore, in view of Proposition 1.3, the solution $v(\cdot)$ is represented by the formula

$$v(t) = G(t)v(0) + \int_0^t G(t-\sigma)[(Bv)(\sigma) + g(\sigma)]\, \mathrm{d}\sigma \; . \tag{2.2}$$

We shall first establish:

Lemma 2.1 *The function $\sigma \in \mathbb{R} \to (Bv)(\sigma) \in X$ is (strongly) almost-periodic.*

Let $B' \in \mathcal{L}(X')$ be the dual operator to B (see (15.14)-(15.16), Chapter II), so that

$$B'x'(x) = x'(Bx) \; , \qquad \forall x \in X, \; \forall x' \in X' \; .$$

We get therefore, $\forall x' \in X'$,

$$x'((Bv)(\sigma)) = (B'x')(v(\sigma)) \; , \qquad \forall \sigma \in \mathbb{R} \; .$$

This equality shows that the function $(Bv)(\cdot)$ is also weakly almost-periodic.

On the other hand, the function $v(\cdot)$ is weakly almost-periodic, hence, in particular, weakly bounded (all numerical almost-periodic functions are bounded on $\mathbb{R}$ - see for instance [18], p. 5, p. 23); we also know that weakly bounded sets are (strongly) bounded (Example 3, Chapter III). Thus the range of $v(\cdot)$: $\{v(t), t \in \mathbb{R}\}$ is a bounded set in X. From the definition of compact operators (Chapter VI) we obtain that the set $\{Bv(t), t \in \mathbb{R}\}$ is conditionally sequentially compact, that is to say, every sequence $\left(Bv(t_n)\right)_{n=1}^{\infty}$ has a convergent subsequence in X.

So the function $(Bv)(\cdot)$ is weakly almost-periodic and has conditionally sequentially compact range; using Chapter 11-3 of [18] we obtain the (strong) almost-periodicity of this function; that completes the proof of our lemma. ■

Proof of Theorem 2.1 - sequel. Thus, we now see that we can write formula (2.2) as

$$v(t) = G(t)v(0) + \int_0^t G(t-\sigma)h(\sigma)\, \mathrm{d}\sigma \tag{2.3}$$

where $h(\cdot)$ is almost-periodic, $\mathbb{R} \to X$ while $v(\cdot)$ is bounded $\mathbb{R} \to X$ (a reflexive B-space). Here, by the strong almost-periodicity of the operator group $G(t)$, we get $t \to G(t)v(0)$ is almost-periodic. Next, we only have to prove that

$$t \to \int_0^t G(t-\sigma)h(\sigma)\,\mathrm{d}\sigma = G(t)\int_0^t G(-\sigma)h(\sigma)\,\mathrm{d}\sigma = w(t) \tag{2.4}$$

is almost-periodic, $\mathbb{R} \to X$. Here $w(\cdot)$ is bounded, $\mathbb{R} \to X$; the function $\sigma \to G(-\sigma)x$ is X-almost-periodic, $\forall x \in X$ (see [18], p. 2, p. 22); the function $\sigma \to G(-\sigma)h(\sigma)$ is also X-almost-periodic (see [18], p. 30); the function $t \to \int_0^t G(-\sigma)h(\sigma)\,\mathrm{d}\sigma = G(-t)w(t)$ is obviously bounded on $\mathbb{R}$ because, $\forall x \in X$, $\sup_{t\in\mathbb{R}} \|G(-t)x\| = C_x < \infty$ and by the uniform boundedness theorem). Finally, the function $t \to \int_0^t G(-\sigma)h(\sigma)\,\mathrm{d}\sigma$ is a bounded indefinite integral of the almost-periodic function $G(-\sigma)h(\sigma)$ in the reflexive B-space X; as such, using Section 2, Chapter 6 of [8], we obtain its almost-periodicity. Therefore $w(t) = G(t)\int_0^t G(-\sigma)h(\sigma)\,\mathrm{d}\sigma$ is itself almost-periodic (again [18], p. 30). ■

Chapter XIII

The well-posed ultraweak Cauchy problem and related semigroups of operators

Introduction

This chapter is closely related to the previous Chapter X. Again we consider ultraweak solutions of the first-order Cauchy problem in Banach spaces which are also well-posed ('continuous dependence on the initial data' is assumed). It is then possible to associate with this Cauchy problem an operator semigroup $T(t)$ and to establish equality between its Laplace transformation and the resolvent operator $(\lambda I - A)^{-1}$ (the differential equation being here $u'(t) = Au(t),\ 0 \le t < \infty,\ u(0) = u_0$).

1. Thus, let X be a Banach space over the complex field $\mathbb{C}$, and then A, $\mathcal{D}(A) \subset X \to X$ be a linear operator with dense domain in X. Thus, its dual operator A', $\mathcal{D}(A') \subset X' \to X'$ is well-defined and the equality

$$x'(Ax) = (A'x')(x)\ , \qquad \forall x \in \mathcal{D}(A),\ \forall x' \in \mathcal{D}(A')\ ,$$

holds true. The class of test-functions $K_{A'}(0,\infty)$ consists, as in Chapter X, of $C^1([0,\infty);X')$ functions, vanishing outside some compact interval $[a,b] \subset (0,\infty)$ (depending on the function); furthermore, if $\varphi'(\cdot) \in K_{A'}(0,\infty)$, then $\varphi'(t) \in \mathcal{D}(A')$, $\forall t \ge 0$, and $(A'\varphi')(\cdot) \in C([0,\infty);X')$.

Again, as in Chapter X, the function $u(\cdot) \in C([0,\infty);X)$ is an ultraweak solution of the differential equation (2.1) iff the integral relation (2.2) holds true. If $u(0) = u_0 \in X$ is given in advance we have the ultraweak Cauchy problem for continuous functions.

We shall next base our discussion on Theorem 1.1 in [21-II] (Chapter XI) and formulate here the following:

Theorem *Let us assume that* $\forall u_0 \in \mathcal{D}(A)$ *there exists a unique solution* $u(\cdot)$ *to the ultraweak Cauchy problem on* $[0,\infty)$*, such that* $u(0) = u_0$ *and* $u(t) \in \mathcal{D}(A)$*,* $\forall t > 0$*. Assume also the pointwise continuous dependence with respect to initial data on* $[0,\infty)$*. It* *follows that there exists an operator semigroup* $U(t),\ t \geq 0 \to \mathcal{L}(X),\ U(0) = I$, *such that* $t \to U(t)x \in C(]0,\infty[;X),\ \forall x \in X$, *and* $u(t) = U(t)u_0,\ t \geq 0$.

Thus, we can say that the semigroup $U(t)$ is strongly continuous on $]0,\infty[$.

Next we also have exponential growth of $U(t)$, precisely

$$\lim_{t\to\infty} \frac{1}{t} \ln\|U(t)\| = w\ , \quad \text{where } -\infty \leq w < +\infty \tag{1.1}$$

(see Proposition 3.1, p. 73 in [21-II]).

We consider the Laplace integral

$$\int_0^\infty e^{-\lambda t} U(t)x \,dt = \int_0^\infty e^{-\lambda t} u(t) \,dt\ , \quad x \in \mathcal{D}(A) \tag{1.2}$$

where $u(t) = U(t)x,\ u(0) = x,\ u(\cdot)$ is an ultraweak solution of $u' - Au = \mathbf{0}$ on $[0,\infty)$.

This is the improper Riemann integral

$$\lim_{R\to\infty} \int_0^R e^{-\lambda t} U(t)x \,dt\ ,$$

which exists if $\operatorname{Re}\lambda > w$; in fact, if $w_1 > w$, $\exists \bar{t} \in [0,\infty)$ such that $\|U(t)\| \leq e^{tw_1}$ for $t \geq \bar{t}$ and therefore we obtain

$$\left\| e^{-\lambda t} U(t)x \right\| = e^{-(\operatorname{Re}\lambda)t} \|U(t)x\| \leq e^{t(w_1 - \operatorname{Re}\lambda)} \|x\|\ , \quad \forall t \geq \bar{t}\ . \tag{1.3}$$

On the other hand, if $\operatorname{Re}\lambda > w$, there exists w_1 such that $w < w_1 < \operatorname{Re}\lambda$. We use (1.3) and see that $\int_0^\infty \left\| e^{-\lambda t} u(t) \right\| dt < \infty$.

Next, we continue in a somewhat restricted situation (as in Chapter X, Theorem 2.1). We thus assume:

Hypothesis *X is a reflexive B-space over* $\mathbb{C}$; *A is linear closed operator in X (with dense domain).*

From the equality $u(t) = U(t)x$ we infer readily that $\|u(t)\| \le \|U(t)\|\,\|x\|$, hence

$$\frac{1}{t}\ln\|u(t)\| \le \frac{1}{t}\ln\|U(t)\| + \frac{1}{t}\ln\|x\| ,$$

whence

$$\limsup_{t\to\infty} \frac{1}{t}\ln\|u(t)\| \le w \tag{1.4}$$

(a property which also appears in Chapter X, (2.3)).

As in the proof of Theorem 2.1, Chapter X, we first note that $\forall w_1 > w$, $\exists R_1 > 0$ such that $\|u(t)\| \le \mathrm{e}^{tw_1}$, $\forall t \ge R_1$.

Then we again consider convolutions $u * \alpha_n$ defined by

$$(u * \alpha_n)(t) = \int_{-\infty}^{\infty} u(s)\alpha_n(t-s)\,\mathrm{d}s$$

which are defined on $[t_0, +\infty)$, where $n \ge n_0$ with $\dfrac{1}{n_0} \le t_0$.

In particular, if $t_0 \ge 1$, $(u * \alpha_n)(t)$ is defined $\forall n \in \mathbb{N}$ and for $t \ge R_1 + 1$, and the estimate (2.6)-X:

$$\|(u * \alpha_n)(t)\| \le C_{w_1}\mathrm{e}^{tw_1} , \qquad t \ge R_1 + 1,\ n \in \mathbb{N}$$

holds true.

Then the improper Riemann integral

$$\int_{t_0}^{\infty} \mathrm{e}^{-\lambda t}(u * \alpha_n)(t)\,\mathrm{d}t , \qquad n \ge n_0,\ t_0 \ge \frac{1}{n_0}$$

is absolutely convergent, $\forall \lambda \in \mathbb{C}$, $\operatorname{Re}\lambda > w$.

On the other hand, as seen in 2-Chapter X, the convolutions $(u*\alpha_n)(\cdot)$ are regular solutions of the (abstract) differential equation

$$\frac{\mathrm{d}}{\mathrm{d}t}(u*\alpha_n)(t) = A(u*\alpha_n)(t)\,, \quad \text{for } t>t_0,\ n>n_0,\ \frac{1}{n_0}<t_0.$$

Continuing as in 3-Chapter X we find that

$$\int_{t_0}^{\infty} \mathrm{e}^{-\lambda t}(u*\alpha_n)(t)\,\mathrm{d}t \in \mathcal{D}(A)$$

(when $\operatorname{Re}\lambda > w,\ n>n_0$), and the equality

$$A\int_{t_0}^{\infty} \mathrm{e}^{-\lambda t}(u*\alpha_n)(t)\,\mathrm{d}t = -\mathrm{e}^{-\lambda t_0}(u*\alpha_n)(t_0) + \lambda\int_{t_0}^{\infty} \mathrm{e}^{-\lambda t}(u*\alpha_n)(t)\,\mathrm{d}t$$

holds true, that is also

$$(\lambda I - A)\int_{t_0}^{\infty} \mathrm{e}^{-\lambda t}(u*\alpha_n)(t)\,\mathrm{d}t = \mathrm{e}^{-\lambda t_0}(u*\alpha_n)(t_0),\ n>n_0,\ \operatorname{Re}\lambda > w\,.$$

Again, we continue as in 3-Chapter X (Proposition 3.2-3.3) to get:

$$\int_{t_0}^{\infty} \mathrm{e}^{-\lambda t}u(t)\,\mathrm{d}t \in \mathcal{D}(A) \quad (\text{for } \operatorname{Re}\lambda > w)$$

and the equality

$$(\lambda I - A)\int_{t_0}^{\infty} \mathrm{e}^{-\lambda t}u(t)\,\mathrm{d}t = \mathrm{e}^{-\lambda t_0}u(t_0)$$

is valid.

Taking the limit as $t_0 \to 0$, using closedness of operator $(\lambda I - A)$, one gets, in a similar way to proposition 3.4-Chapter X, that

$$\int_0^\infty e^{-\lambda t} u(t)\, dt \in \mathcal{D}(A) \quad \text{for} \quad \operatorname{Re}\lambda > w$$

and the equality

$$(\lambda I - A)\int_0^\infty e^{-\lambda t} u(t)\, dt = u(0) = x \tag{1.5}$$

holds true.

This is true $\forall x \in \mathcal{D}(A)$, so that we can infer that

$$\mathcal{D}(A) \subset (\lambda I - A)(\mathcal{D}(A)) \ .$$

Note also Proposition 3.2, p. 73 in [21-II], which implies that for $\lambda \in \mathbb{C}$, $\operatorname{Re}\lambda > w$, the operator $(\lambda I - A)^{-1}$ exists.

Then, from (1.5) we readily derive the equality

$$\int_0^\infty e^{-\lambda t} U(t)x\, dt = (\lambda I - A)^{-1} x \ , \quad \forall x \in \mathcal{D}(A), \ \operatorname{Re}\lambda > w, \tag{1.6}$$

a well-known connection between the semigroup $U(t)$ and the resolvent operator of A, which is here established in an ultraweak Cauchy problem setting.

Chapter XIV

Asymptotic result for ultraweak differential inequalities in Hilbert space

Introduction

In this section we present, following [16], a result concerning *ultraweak* differential inequalities in Hilbert space; their *strong* formulation is $\|u'(t)-Au(t)\| \le \phi(t)\|u(t)\|,\ t \ge 0$ and, as seen in [21-II, p. 62], solutions on $[0,\infty)$ which decrease faster than any exponential at $+\infty$ are only the zero solutions.

1. Let H be a Hilbert space and let A be a linear closed operator, $\mathcal{D}(A) \subset H \to H$, where the domain $\mathcal{D}(A)$ is dense in H.

Let A^* be the (hilbertian) adjoint to A, which is well defined on the set

$$\mathcal{D}(A^*) = \{h \in H, \text{ such that } \exists h^* \in H, \text{ with the property } (Ak,h) = (k,h^*),\ \forall k \in \mathcal{D}(A)\}$$

via the formula $A^*h = h^*$.

Thus, $(Ah,k) = (h, A^*k)$ holds, $\forall h \in \mathcal{D}(A)$, $\forall k \in \mathcal{D}(A^*)$ (see Chapter II-5).

From Theorem 11.1, Chapter II, we also derive that $\mathcal{D}(A^*)$ is also dense in H.

Define the class of test-functions $K_{A^*}((0,\infty))$ which consists of functions $\varphi(\cdot)$, $[0,\infty) \to \mathcal{D}(A^*)$, such that $\varphi(\cdot) \in C^1([0,\infty);H)$, $(A^*\varphi)(\cdot) \in C([0,\infty),H)$, and $\varphi(t) = \mathbf{0}$ outside a compact interval $[a,b] \subset (0,\infty)$, (depending on $\varphi(\cdot)$).

Our aim is to demonstrate:

Theorem. *Let $u(\cdot), f(\cdot) \in C([0,\infty);H)$ such that*

$$\int_0^\infty (u(t), \varphi'(t) + (A^*\varphi)(t))\,\mathrm{d}t = -\int_0^\infty (f(t), \varphi(t))\,\mathrm{d}t \tag{1.1}$$

holds, $\forall \varphi(\cdot) \in K_{A^*}((0,\infty))$.

Assume also that on a sequence of vertical lines in the complex plane: $(\operatorname{Re}\lambda = \sigma_n)_{n=1}^\infty$, where $\sigma_n \to -\infty$, the resolvent operator

$$(\sigma_n + \mathrm{i}\tau - A)^{-1} \in \mathcal{L}(H)\,, \qquad \forall n \in \mathbb{N},\ \forall \tau \in \mathbb{R}$$

and

$$\left\| (\sigma_n + \mathrm{i}\tau - A)^{-1} \right\|_{\mathcal{L}(H)} \le M\,, \qquad \forall n \in \mathbb{N},\ \forall \tau \in \mathbb{R}\,. \tag{1.2}$$

Under these assumptions, if

$$\|f(t)\| \le \alpha(t)\|u(t)\|,\ 0 \le t < \infty \quad \text{where} \quad \alpha(t) \le c < \frac{1}{M} \tag{1.3}$$

on $[0,\infty)$, *and if, finally,*

$$\sup_{t \ge 0}\ \mathrm{e}^{-at}\|u(t)\| < \infty\,, \qquad \forall a \in \mathbb{R} \tag{1.4}$$

it follows that $u(t) = \mathbf{0},\ \forall t \ge 0$.

The proof is quite long and involved; it makes essential use of the mollification procedure. Our first step is to transform (1.1) into a relation on the whole real line.

Lemma 1 *Let* $\zeta(\cdot)$ *be a* $C^1(\mathbb{R})$ *function,* $=0$ *on* $]-\infty,0]$, $=1$ *on* $[t_0,\infty)$ *with some* $t_0 > 0$, *increasing on* $[0,t_0]$. *Let* $v(\cdot)$ *be defined by* $v(t) = \mathrm{e}^{at}\zeta(t)u(t),\ t \ge 0,\ = \mathbf{0},\ t < 0$, *where a is a real number. Then the integral identity*

$$\int_{\mathbb{R}} (v(t),\ \psi'(t) + (A^*\psi)(t))\,\mathrm{d}t = -\int_{\mathbb{R}} (h(t),\ \psi(t))\,\mathrm{d}t \tag{1.5}$$

holds, where $\psi \in K_{A^*}(\mathbb{R})$ *and* $h(t) = \mathrm{e}^{at}\zeta(t)f(t) + \mathrm{e}^{at}\zeta'(t)u(t) + av(t)$, *for* $t \ge 0$, $= \mathbf{0}$ *for* $t < 0$.

Here $K_{A^*}(\mathbb{R})$ is the class of test-functions defined, for instance, in XI-1.

The demonstration for this lemma goes on as follows.

We have

$$\int_{\mathbb{R}} (v(t),\ \psi'(t))\,\mathrm{d}t = \int_0^\infty \left(\mathrm{e}^{at}\zeta(t)u(t),\ \psi'(t)\right)\mathrm{d}t = \int_0^\infty \left(u(t),\ \mathrm{e}^{at}\zeta(t)\psi'(t)\right)\mathrm{d}t\,.$$

On the other hand, the obvious identity

$$e^{at}\zeta(t)\psi'(t) = \left(e^{at}\zeta(t)\psi(t)\right)' - \left(e^{at}\zeta(t)\right)'\psi(t)$$

holds true, whence

$$\int_{\mathbb{R}} (v(t),\ \psi'(t))\,dt = \int_0^\infty \left(u(t),\ \left(e^{at}\zeta(t)\psi(t)\right)' \right) dt - \int_0^\infty \left(u(t), \left(e^{at}\zeta(t)\right)'\psi(t) \right). \tag{1.6}$$

Denote now $e^{at}\zeta(t)\psi(t) = \varphi(t)$ and then consider, $\forall \varepsilon > 0$, a scalar-valued function $v_\varepsilon(\cdot)$ such that

$$v_\varepsilon(t) = 0 \text{ for } 0 \le t \le \varepsilon, \quad v_\varepsilon(t) = 1 \text{ for } t > 3\varepsilon, \quad v_\varepsilon(\cdot) \in C^1([0,\infty)), \quad |v_\varepsilon(t)| \le 1,$$

$$|v'_\varepsilon(t)| \le \frac{C}{\varepsilon}, \quad \forall t \ge 0,\ \forall \varepsilon > 0.$$

Similar functions have been explicitly constructed in [21-II]; let us give here details for the present function $v_\varepsilon(\cdot)$; take $\Lambda(t) = 0,\ -\infty \le t \le 2\varepsilon,\ \Lambda(t) = 1,\ 2\varepsilon < t < \infty$; then take $K(\cdot) \in C_0^\infty([-1,1]),\ K(t) \ge 0,\ \forall t \in \mathbb{R},\ \int_{\mathbb{R}} K(s)\,ds = 1$; define now $K_d(t) = K\left(\frac{t}{d}\right),\ \forall t \in \mathbb{R}$, and then

$$\mu_d(t) = \frac{1}{d}\int_{\mathbb{R}} \Lambda(s) K_d(t-s)\,ds = \frac{1}{d}\int_{t-d}^{t+d} \Lambda(s) K_d(t-s)\,ds\ .$$

One sees that $\mu_d(\cdot) \in C^\infty(\mathbb{R}),\ \mu_d(t) = 0$ for $-\infty \le t \le 2\varepsilon - d,\ \mu_d(t) = 1$ for $2\varepsilon + d \le t < \infty$; put now $v_\varepsilon(\cdot) = \mu_\varepsilon(\cdot)$; thus $v_\varepsilon(t) \in C^\infty(\mathbb{R}),\ v_\varepsilon(t) = 0$ for $-\infty \le t \le \varepsilon,\ v_\varepsilon(t) = 1$ for $3\varepsilon < t < \infty$. To estimate $v'_\varepsilon(\cdot)$ we first get

$$v_\varepsilon(t) = \mu_\varepsilon(t) = \frac{1}{\varepsilon}\int_{\mathbb{R}} \Lambda(s) K\left(\frac{t-s}{\varepsilon}\right) ds\ ,$$

whence

$$|v_\varepsilon(t)| \le \frac{1}{\varepsilon}\int_{\mathbb{R}} K\left(\frac{t-s}{\varepsilon}\right) ds = 1$$

while

$$v_\varepsilon'(t) = \frac{1}{\varepsilon}\int_{\mathbb{R}} \Lambda(s)\cdot\frac{1}{\varepsilon}K'\left(\frac{t-s}{\varepsilon}\right)\mathrm{d}s$$

and then, $\forall t \in \mathbb{R}$

$$\left|v_\varepsilon'(t)\right| \le \frac{1}{\varepsilon^2}\int_{\mathbb{R}}\left|K'\left(\frac{t-s}{\varepsilon}\right)\right|\mathrm{d}s = \frac{1}{\varepsilon^2}\int_{\mathbb{R}}|K'(u)|\varepsilon\,\mathrm{d}u = \frac{c}{\varepsilon}\,.$$

Afterwards we note that the product function: $t \to v_\varepsilon(t)\varphi(t)$, belongs to $K_{A^*}(]0,\infty))$ so that, introducing in (1.1), we get, $\forall\varepsilon > 0$

$$\int_0^\infty \left(u(t),\ (v_\varepsilon\varphi)'(t)\right)\mathrm{d}t = -\int_0^\infty \left(u(t),\ A^*(v_\varepsilon\varphi)(t)\right)\mathrm{d}t - \int_0^\infty \left(f(t),\ (v_\varepsilon\varphi)(t)\right)\mathrm{d}t\,. \qquad (1.7)$$

Next, let us note that

$$\int_0^\infty \left(u(t),\ (v_\varepsilon\varphi)'(t)\right)\mathrm{d}t = \int_\varepsilon^{3\varepsilon}\left(u(t),\ v_\varepsilon(t)\varphi'(t)\right)\mathrm{d}t + \int_\varepsilon^{3\varepsilon}\left(u(t),\ v_\varepsilon'(t)\varphi(t)\right)\mathrm{d}t$$

$$+\int_{3\varepsilon}^\infty (u(t),\ \varphi'(t))\,\mathrm{d}t\,.$$

Now

$$\left|\int_\varepsilon^{3\varepsilon}\left(u(t),\ v_\varepsilon(t)\varphi'(t)\right)\mathrm{d}t\right| \le 2\varepsilon \sup_{\varepsilon\le t\le 3\varepsilon}\|u(t)\|\,\|\varphi'(t)\|\cdot c \to 0 \quad \text{as} \quad \varepsilon\to 0\,;$$

Also,

$$\lim_{\varepsilon\to 0}\int_{3\varepsilon}^\infty (u(t),\ \varphi'(t))\,\mathrm{d}t = \int_0^\infty (u(t),\ \varphi'(t))\,\mathrm{d}t$$

and

$$\left|\int_\varepsilon^{3\varepsilon}\left(u(t),\ v_\varepsilon'(t)\varphi(t)\right)\mathrm{d}t\right| \le 2\varepsilon\cdot\frac{c}{\varepsilon}\sup_{\varepsilon\le t\le 3\varepsilon}\|u(t)\|\,\|\varphi(t)\| = c_1\sup_{\varepsilon\le t\le 3\varepsilon}\|u(t)\|\,\|\varphi(t)\|\,.$$

As $u(t)\to u(0)$ for $t\to 0$ while $\varphi(t)\to\varphi(0)=\mathbf{0}$ for $t\to 0$ ($\varphi(0)=\zeta(0)\psi(0)=\mathbf{0}$, as $\zeta(0)=0$), one gets

$$\lim_{\varepsilon\to 0} \int_{\varepsilon}^{3\varepsilon} \big(u(t),\ v'_{\varepsilon}(t)\varphi(t)\big)\,\mathrm{d}t = 0 .$$

Summing these limits, we can infer that

$$\lim_{\varepsilon\to 0} \int_0^{\infty} \Big(u(t),\ \big(v_{\varepsilon}\varphi\big)'(t) \Big)\,\mathrm{d}t = \int_0^{\infty} (u(t),\ \varphi'(t))\,\mathrm{d}t .$$

Looking now at the right-hand side of (1.7), we first see that

$$-\int_0^{\infty} \big(u(t),\ A^*\big(v_{\varepsilon}\varphi\big)(t)\big)\,\mathrm{d}t = -\int_{\varepsilon}^{3\varepsilon} \big(u(t),\ \big(A^*\varphi\big)(t)\big)v_{\varepsilon}(t)\,\mathrm{d}t - \int_{3\varepsilon}^{\infty} (u(t),\ (A^*\varphi)(t))\,\mathrm{d}t$$

which has the obvious limit (for $\varepsilon \to 0$)

$$\lim_{\varepsilon\to 0}\left(-\int_0^{\infty} \big(u(t),\ A^*\big(v_{\varepsilon}\varphi\big)(t)\big)\,\mathrm{d}t \right) = -\int_0^{\infty} (u(t),\ (A^*\varphi)(t))\,\mathrm{d}t .$$

Similarly, or by use of Lebesgue's dominated convergence theorem, one gets

$$\lim_{\varepsilon\to 0}\left(-\int_0^{\infty} \big(f(t),\ \big(v_{\varepsilon}\varphi\big)(t)\big)\,\mathrm{d}t \right) = -\int_0^{\infty} (f(t),\ \varphi(t))\,\mathrm{d}t .$$

Therefore we get, as $\varepsilon \to 0$, from (1.7), the equality

$$\int_0^{\infty} (u(t),\ \varphi'(t))\,\mathrm{d}t = \int_0^{\infty} (u(t),\ (A^*\varphi)(t))\,\mathrm{d}t \quad \int_0^{\infty} (f(t),\ \varphi(t))\,\mathrm{d}t \tag{1.8}$$

where $\varphi(t) = \mathrm{e}^{at}\zeta(t)\psi(t)$, and $\psi \in K_{A^*}(-\infty,\infty)$.

Actually (1.8) is the same as

$$\int_0^{\infty} \Big(u(t),\ \big(\mathrm{e}^{at}\zeta(t)\psi(t)\big)' \Big)\,\mathrm{d}t = -\int_0^{\infty} \big(u(t),\ \mathrm{e}^{at}\zeta(t)(A^*\psi)(t)\big)\,\mathrm{d}t - \int_0^{\infty} \big(f(t),\ \mathrm{e}^{at}\zeta(t)\psi(t)\big)\,\mathrm{d}t$$

which, in view of (1.6), can be written in the form

$$\int_{-\infty}^{\infty}(v(t),\ \psi'(t))\,\mathrm{d}t+\int_{0}^{\infty}\left(u(t),\ \left(\mathrm{e}^{at}\zeta(t)\right)'\psi(t)\right)\mathrm{d}t=-\int_{0}^{\infty}\left(u(t),\ \mathrm{e}^{at}\zeta(t)(A^{*}\psi)(t)\right)\mathrm{d}t$$

$$-\int_{0}^{\infty}\left(f(t),\ \mathrm{e}^{at}\zeta(t)\psi(t)\right)\mathrm{d}t=-\int_{-\infty}^{\infty}(v(t),\ A^{*}\psi(t))\,\mathrm{d}t-\int_{-\infty}^{\infty}\left(\mathrm{e}^{at}\zeta(t)f(t),\ \psi(t)\right)\mathrm{d}t$$

and also

$$\int_{-\infty}^{\infty}(v(t),\ \psi'(t)+(A^{*}\psi)(t))\,\mathrm{d}t=-\int_{-\infty}^{\infty}\left(\mathrm{e}^{at}\zeta(t)f(t),\ \psi(t)\right)\mathrm{d}t$$

$$-\int_{0}^{\infty}\left(u(t),\ \left(\mathrm{e}^{at}\zeta(t)\right)'\psi(t)\right)\mathrm{d}t=-\int_{-\infty}^{\infty}\left(\mathrm{e}^{at}\zeta(t)f(t),\ \psi(t)\right)\mathrm{d}t$$

$$-\int_{0}^{\infty}\left(\mathrm{e}^{at}\zeta'(t)u(t)+a\mathrm{e}^{at}\zeta(t)u(t),\ \psi(t)\right)\mathrm{d}t$$

which is in fact (1.5).

Note that the function h which equals

$$\mathrm{e}^{at}\zeta(t)f(t)+\mathrm{e}^{at}\zeta'(t)u(t)+av(t),\ t\geq 0,\ \text{and } \mathbf{0} \text{ for } t<0, \text{ belongs to } C(\mathbb{R};H)\ . \quad (1.9)$$

2. In this section we start by regularizing the solutions of (1.5); precisely, let us consider a sequence of scalar-valued, non-negative $C^{1}(\mathbb{R})$-functions, $\left(\alpha_{n}(\cdot)\right)_{1}^{\infty}$, vanishing for $|t|\geq\frac{1}{n}$, such that $\int_{\mathbb{R}}\alpha_{n}(s)\,\mathrm{d}s=1,\ \forall n=1,2,\ldots$; next, we have the convolutions

$$(v*\alpha_{n})(t)=\int_{\mathbb{R}}v(s)\alpha_{n}(t-s)\,\mathrm{d}s=\int_{|t-s|\leq\frac{1}{n}}v(s)\alpha_{n}(t-s)\,\mathrm{d}s\ ,$$

and $h*\alpha_{n}$ defined by similar formula.

One sees immediately that the derivative $\left(v*\alpha_{n}\right)'(t)$ equals

$$\int_{\mathbb{R}} v(s)\alpha_n'(t-s)\,\mathrm{d}s = \int_{t-\frac{1}{n}}^{t+\frac{1}{n}} v(s)\alpha_n'(t-s)\,\mathrm{d}s\ .$$

We show next the norm-integrability on the real line of the functions $(v*\alpha_n)'(\cdot)$ and $(h*\alpha_n)(\cdot)$, as well as of $(v*\alpha_n)(t)$.

First we obviously have the estimate for $v(\cdot)$:

$$\|v(t)\| = \mathrm{e}^{at}\zeta(t)\|u(t)\|\ , \quad t \geq 0, \quad = 0 \quad \text{for} \quad t < 0\ .$$

On the other hand, from (1.4) we infer that, $\forall b \in \mathbb{R}$, $\exists N_b \in \mathbb{R}^+$, such that

$$\|u(t)\| \leq N_b \mathrm{e}^{bt}, \quad \forall t \geq 0\,.$$

Hence

$$\|v(t)\| \leq N_b \mathrm{e}^{(a+b)t}, \quad \forall t \in \mathbb{R}\,.$$

We shall choose, given $a \in \mathbb{R}$ as in (1.5), any $b \in \mathbb{R}$ such that $a+b=\beta<0$, and now we have the estimate $\|v(t)\| \leq N_b \mathrm{e}^{\beta t}$, $\forall t \in \mathbb{R}$. It follows accordingly that

$$\left\|(v*\alpha_n)'(t)\right\| \leq \int_{t-\frac{1}{n}}^{t+\frac{1}{n}} \|v(s)\|\,|\alpha_n'(t-s)|\,\mathrm{d}s = \int_{-\frac{1}{n}}^{\frac{1}{n}} \|v(t-\sigma)\|\,|\alpha_n'(\sigma)|\,\mathrm{d}\sigma$$

$$\leq N_b \int_{-\frac{1}{n}}^{\frac{1}{n}} \mathrm{e}^{\beta(t-\sigma)}|\alpha_n'(\sigma)|\,\mathrm{d}\sigma = N_b \mathrm{e}^{\beta t} \int_{-\frac{1}{n}}^{\frac{1}{n}} \mathrm{e}^{|\beta|\sigma}|\alpha_n'(\sigma)|\,\mathrm{d}\sigma$$

$$\leq N_b \mathrm{e}^{\beta t}\mathrm{e}^{|\beta|\frac{1}{n}} \int_{-\frac{1}{n}}^{\frac{1}{n}} |\alpha_n'(\sigma)|\,\mathrm{d}\sigma = C_n \mathrm{e}^{\beta t}\ , \quad \forall t \in \mathbb{R}\,.$$

Furthermore we note that $(v*\alpha_n)'(t) = \mathbf{0}$ for $t < -\dfrac{1}{n}$. Therefore, $(v*\alpha_n)$ is absolutely integrable on the real line.

Let us estimate next the function $h(\cdot)$ and then $(h*\alpha_n)(\cdot)$; remember that $h(t)=\mathbf{0}$ for $t<0$, and $h(t)=e^{at}\zeta(t)f(t)+e^{at}\zeta'(t)u(t)+av(t)$, for $t\ge 0$, whence we get

$$\|h(t)\|\le e^{at}\|f(t)\|+e^{at}|\zeta'(t)|\,\|u(t)\|+|a|\,\|v(t)\|\,,\qquad t\ge 0\ .$$

Actually, from (1.3) we get, for $t\ge 0$

$$\|f(t)\|\le \alpha(t)\|u(t)\|\le c\|u(t)\|\le CN_b e^{bt}\,,\qquad t\ge 0\ .$$

Also, $\zeta'(t)=0$ for $t\ge t_0$ and $\|av(t)\|\le |a|N_b e^{\beta t}$, $t\ge 0$. Obviously too, $\exists c_1>0$ such that $e^{at}|\zeta'(t)|\,\|u(t)\|\le c_1 e^{\beta t}$, $t\ge 0$.

Consequently we immediately get that

$$\|h(t)\|\le Ce^{\beta t}\,,\quad t\ge 0\quad \text{and}\quad h(t)=\mathbf{0}\,,\quad \text{for } t<0.$$

We obtain therefore, for the convolution $h*\alpha_n$, the estimate

$$\begin{aligned}\|(h*\alpha_n)(t)\| &\le \int_{-\frac{1}{n}}^{1/n}\|h(t-\sigma)\|\alpha_n(\sigma)\,d\sigma\\ &\le C_1\int_{-\frac{1}{n}}^{\frac{1}{n}} e^{\beta(t-\sigma)}\alpha_n(\sigma)\,d\sigma\\ &\le C_1 e^{\beta t}e^{|\beta|\frac{1}{n}}\\ &= C_2 e^{\beta t}\,,\qquad \forall t\in\mathbb{R}\end{aligned}$$

and we also know that $(h*\alpha_n)(t)=\mathbf{0}$ for $t<-\dfrac{1}{n}$. Again, therefore, $h*\alpha_n$ is absolutely integrable on $\mathbb{R}$.

A similar computation works for $v*\alpha_n$, which is also absolutely integrable on $\mathbb{R}$.

We shall next establish that $(v*\alpha_n)(t)\in\mathcal{D}(A)$, $\forall t\in\mathbb{R}$ and that the equality

$$(v*\alpha_n)'(t)=A(v*\alpha_n)(t)+(h*\alpha_n)(t)\,,\qquad t\in\mathbb{R} \tag{2.1}$$

holds true, $\forall n = 1,2,\dots$.

This is in fact a consequence of the integral identity (1.5); the proof is similar to that of (1.9), Chapter XI.

Thus, take in (1.5), test-functions $\psi_{n,t_0}(t) = \alpha_n(t_0 - t)k$, where $k \in \mathcal{D}(A^*)$. One obtains

$$\int_{\mathbb{R}} \left(v(t),\ -\alpha_n'(t_0 - t)k + (A^* k)\alpha_n(t_0 - t)\right) \mathrm{d}t = -\int_{\mathbb{R}} \left(h(t),\ \alpha_n(t_0 - t)k\right) \mathrm{d}t\ ,$$

$$\forall n \in \mathbb{N},\ \forall k \in \mathcal{D}(A^*),\ \forall t_0 \in \mathbb{R}\,.$$

This can also be written in the form

$$\left(\int_{\mathbb{R}} \alpha_n(t_0 - t)v(t)\,\mathrm{d}t,\ A^* k\right) + \left(\int_{\mathbb{R}} \alpha_n(t_0 - t)h(t),\ k\right)\mathrm{d}t = \left(\int_{\mathbb{R}} \alpha_n'(t_0 - t)v(t)\,\mathrm{d}t,\ k\right),$$

$$\forall k \in \mathcal{D}(A^*),\ \forall t_0 \in \mathbb{R}$$

and this is turn implies that

$$\int_{\mathbb{R}} \alpha_n(t_0 - t)v(t)\,\mathrm{d}t \in \mathcal{D}(A^{**})$$

and

$$A^{**} \int_{\mathbb{R}} \alpha_n(t_0 - t)v(t)\,\mathrm{d}t = \int_{\mathbb{R}} \alpha_n'(t_0 - t)v(t)\,\mathrm{d}t - \int_{\mathbb{R}} \alpha_n(t_0 - t)h(t)\,\mathrm{d}t\ , \quad \forall t_0 \in \mathbb{R},\ n \in \mathbb{N}.$$

Now as A is closed, we know from previous Chapters that $A^{**} = A$. We get therefore the equality we were looking for

$$A(v * \alpha_n)(t_0) = (v * \alpha_n)'(t_0) - (h * \alpha_n)(t_0)\,, \quad \forall t_0 \in \mathbb{R},\ \forall n \in \mathbb{N}\,.$$

This has in particular the consequence: $A(v * \alpha_n)(\cdot)$ is also absolutely integrable on the real line.

3. Once we have at our disposition equality (2.1) and various estimates proved on p. 193 we can proceed to 'Fourier-transform' things in a similar way to [21-II, p. 63-5].

Let us multiply equality (2.1) by $\dfrac{1}{\sqrt{2\pi}}\mathrm{e}^{-\mathrm{i}\tau t}$ $(\tau \in \mathbb{R},\ \mathrm{i} = \sqrt{-1})$ and then integrate on $[-r, r]$, $r > 0$. We get

$$\frac{1}{\sqrt{2\pi}}\int_{-r}^{r} e^{-i\tau t}\left(v*\alpha_n\right)'(t)\,dt = \frac{1}{\sqrt{2\pi}}\int_{-r}^{r} e^{-i\tau t}A\left(v*\alpha_n\right)(t)\,dt + \frac{1}{\sqrt{2\pi}}\int_{-r}^{r} e^{-i\tau t}\left(h*\alpha_n\right)(t)\,dt .$$

Next, we use integration by parts in the left-hand side of this equality to obtain:

$$\frac{1}{\sqrt{2\pi}}\int_{-r}^{r} e^{-i\tau t}\left(v*\alpha_n\right)'(t)\,dt = \frac{1}{\sqrt{2\pi}}\left[e^{-i\tau r}\left(v*\alpha_n\right)(r)\right] + \frac{i\tau}{\sqrt{2\pi}}\int_{-r}^{r}\left(v*\alpha_n\right)(t)e^{-i\tau t}\,dt$$

(as $\left(v*\alpha_n\right)(-r)=0$ for $r<-\frac{1}{n}$).

As noted previously, $\left\|\left(v*\alpha_n\right)(t)\right\|\le ce^{\beta t}$ $(\beta<0)$, $\forall t\in\mathbb{R}$, so that $\left(v*\alpha_n\right)(r)\to\mathbf{0}$ as $r\to+\infty$. It follows that

$$\frac{1}{\sqrt{2\pi}}\int_{-\infty}^{\infty} e^{-i\tau t}\left(v*\alpha_n\right)'(t)\,dt = i\tau\frac{1}{\sqrt{2\pi}}\int_{-\infty}^{\infty} e^{-i\tau t}\left(v*\alpha_n\right)(t)\,dt \tag{3.1}$$

where both improper integrals are absolutely convergent.

We next use the commutativity of closed operators with improper integrals (precisely Proposition 3.1-Chapter X), extended slightly to the case $a=-\infty$, $b=+\infty$. The closed operator here is A. We know that

$$\frac{1}{\sqrt{2\pi}}\int_{-\infty}^{\infty}\left(v*\alpha_n\right)(t)\,dt \quad\text{and}\quad \frac{1}{\sqrt{2\pi}}\int_{-\infty}^{\infty}A\left(v*\alpha_n\right)(t)\,dt$$

are both absolutely convergent (improper) integrals; we infer accordingly

$$\frac{1}{\sqrt{2\pi}}\int_{-\infty}^{\infty}\left(v*\alpha_n\right)(t)\,dt\in\mathcal{D}(A)$$

and

$$A\frac{1}{\sqrt{2\pi}}\int_{-\infty}^{\infty}\left(v*\alpha_n\right)(t)\,dt = \frac{1}{\sqrt{2\pi}}\int_{-\infty}^{\infty}A\left(v*\alpha_n\right)(t)\,dt$$

and therefore

$$\frac{1}{\sqrt{2\pi}}\int_{-\infty}^{\infty} \mathrm{e}^{-\mathrm{i}\tau t}(v*\alpha_n)(t)\,\mathrm{d}t \in \mathcal{D}(A)\,,$$

$$A\frac{1}{\sqrt{2\pi}}\int_{-\infty}^{\infty} \mathrm{e}^{-\mathrm{i}\tau t}(v*\alpha_n)(t)\,\mathrm{d}t = \frac{1}{\sqrt{2\pi}}\int_{-\infty}^{\infty} \mathrm{e}^{-\mathrm{i}\tau t}A(v*\alpha_n)(t)\,\mathrm{d}t\,. \tag{3.2}$$

It follows that

$$\mathrm{i}\tau\frac{1}{\sqrt{2\pi}}\int_{-\infty}^{\infty} \mathrm{e}^{-\mathrm{i}\tau t}(v*\alpha_n)(t)\,\mathrm{d}t = A\frac{1}{\sqrt{2\pi}}\int_{-\infty}^{\infty} \mathrm{e}^{-\mathrm{i}\tau t}(v*\alpha_n)(t)\,\mathrm{d}t$$

$$+\frac{1}{\sqrt{2\pi}}\int_{-\infty}^{\infty} \mathrm{e}^{-\mathrm{i}\tau t}(h*\alpha_n)(t)\,\mathrm{d}t\,. \tag{3.3}$$

If we use the standard notation

$$\hat{\xi}(\tau) = \frac{1}{\sqrt{2\pi}}\int_{-\infty}^{\infty} \mathrm{e}^{-\mathrm{i}\tau t}\xi(t)\,\mathrm{d}t$$

for the Fourier transformation of absolutely integrable functions $\xi(\cdot)$, the above equality (3.3) can be written as

$$(\mathrm{i}\tau - A)(v*\alpha_n)^{\wedge}(\tau) = (h*\alpha_n)^{\wedge}(\tau)\,, \qquad \tau\in\mathbb{R},\ n\in\mathbb{N}\,. \tag{3.4}$$

Let us return now to the definition of the function $h(\cdot)$ (1.9). It follows that

$$h(t) = g(t) + av(t)\,,\quad t\geq 0, \qquad h(t) = \mathbf{0}\,,\quad t<0,$$

where

$$g(t) = \mathrm{e}^{at}\zeta(t)f(t) + \mathrm{e}^{at}\zeta'(t)u(t)\,.$$

Then, obviously

$$(h*\alpha_n)(t) = (g*\alpha_n)(t) + a(v*\alpha_n)(t)$$

and consequently, by simple linearity properties of the Fourier transform,

$$(h*\alpha_n)^\wedge(\tau)=(g*\alpha_n)^\wedge(\tau)+a(v*\alpha_n)^\wedge(\tau)\,,\qquad \tau\in\mathbb{R},\ n\in\mathbb{N}\,. \tag{3.5}$$

Substituting in (3.4) one obtains

$$(\mathrm{i}\tau-A)(v*\alpha_n)^\wedge(\tau)=a(v*\alpha_n)^\wedge(\tau)+(g*\alpha_n)^\wedge(\tau)\,,$$

that is

$$(\mathrm{i}\tau-a-A)(v*\alpha_n)^\wedge(\tau)=(g*\alpha_n)^\wedge(\tau)$$

(here a is any real number). Take now $a=-\sigma_n$ (the sequence (σ_n) appears in the statement of the Theorem). We obtain the representation formula

$$(v*\alpha_n)^\wedge(\tau)=(\sigma_n+\mathrm{i}\tau-A)^{-1}(g*\alpha_n)^\wedge(\tau)\,,\qquad n\in\mathbb{N},\ \tau\in\mathbb{R}\,. \tag{3.6}$$

Note now, for any absolutely integrable function $\xi(\cdot)$, the relation $(\xi*\alpha_n)^\wedge(\tau)=\hat{\xi}(\tau)\cdot\hat{\alpha}_n(\tau)(\sqrt{2\pi})$: in fact, we have

$$\frac{1}{\sqrt{2\pi}}\int_{-\infty}^{\infty}\mathrm{e}^{-\mathrm{i}\tau t}\left(\int_{\mathbb{R}}\xi(s)\alpha_n(t-s)\,\mathrm{d}s\right)\mathrm{d}t=\frac{1}{\sqrt{2\pi}}\int_{\mathbb{R}}\mathrm{e}^{-\mathrm{i}\tau(t-s)}\left(\mathrm{e}^{-\mathrm{i}\tau s}\int_{\mathbb{R}}\xi(s)\alpha_n(t-s)\,\mathrm{d}s\right)\mathrm{d}t$$

$$=\frac{1}{\sqrt{2\pi}}\int_{\mathbb{R}}\int_{\mathbb{R}}\mathrm{e}^{-\mathrm{i}\tau(t-s)}\mathrm{e}^{-\mathrm{i}\tau s}\alpha_n(t-s)\xi(s)\,\mathrm{d}t\,\mathrm{d}s$$

$$=\int_{\mathbb{R}}\left[\frac{1}{\sqrt{2\pi}}\int_{\mathbb{R}}\mathrm{e}^{-\mathrm{i}\tau(t-s)}\alpha_n(t-s)\,\mathrm{d}t\right]\mathrm{e}^{-\mathrm{i}\tau s}\xi(s)\,\mathrm{d}s$$

$$=\hat{\alpha}_n(\tau)\sqrt{2\pi}\;\hat{\xi}(\tau)\,.$$

Accordingly, (3.6) becomes

$$\hat{v}(\tau)\hat{\alpha}_n(\tau)=(\sigma_n+\mathrm{i}\tau-A)^{-1}\hat{g}(\tau)\hat{\alpha}_n(\tau)\,,\qquad \tau\in\mathbb{R},\ n\in\mathbb{N}\,. \tag{3.7}$$

Consider now a special sequence of functions $\alpha_n(\cdot)$, which is constructed as follows. Take $\alpha(\cdot)\in C^1(\mathbb{R})$, $\alpha(t)\ge 0$, $\forall t\in\mathbb{R}$, $\alpha(t)=0$ for $|t|\ge 1$, with $\int_{\mathbb{R}}\alpha(s)\,\mathrm{d}s=1$. Then put

$\alpha_n(t) = n\alpha(nt)$. We see that $\alpha_n(\cdot) \in C^1(\mathbb{R})$, $\alpha_n(t) \geq 0$, $\forall t \in \mathbb{R}$, $\alpha_n(t) = 0$ for $|t| \geq \dfrac{1}{n}$, $\int\limits_{\mathbb{R}} \alpha_n(s)\,\mathrm{d}s = \int\limits_{-\infty}^{\infty} n\alpha(ns)\,\mathrm{d}s = \int\limits_{\mathbb{R}} \alpha(\sigma)\,\mathrm{d}\sigma = 1$.

Furthermore, we see that

$$\hat{\alpha}_n(\tau) = \frac{1}{\sqrt{2\pi}} \int\limits_{\mathbb{R}} \mathrm{e}^{-\mathrm{i}\tau t} n\alpha(nt)\,\mathrm{d}t = \frac{1}{\sqrt{2\pi}} \int\limits_{\mathbb{R}} \mathrm{e}^{-\mathrm{i}\tau \frac{\sigma}{n}} \alpha(\sigma)\,\mathrm{d}\sigma = \hat{\alpha}\left(\frac{\tau}{n}\right).$$

Turning back to (3.7) and using (1.2), one gets the estimate

$$\left|\hat{\alpha}\left(\frac{\tau}{n}\right)\right| \|\hat{v}(\tau)\| \leq M \left|\hat{\alpha}\left(\frac{\tau}{n}\right)\right| \|\hat{g}(\tau)\|, \qquad \tau \in \mathbb{R},\ n = 1,2,\ldots. \tag{3.8}$$

If $n \to \infty$, $\hat{\alpha}\left(\dfrac{\tau}{n}\right) \to \hat{\alpha}(0) = \dfrac{1}{\sqrt{2\pi}}$ so that from (3.8) it follows that

$$\|\hat{v}(\tau)\| \leq M \|\hat{g}(\tau)\|, \qquad \forall \tau \in \mathbb{R}.$$

Once we are here we can terminate as in [21-I, p. 65-7].

This ends the proof of the Theorem.

Chapter XV

Almost-periodic solutions II

Introduction

The purpose of this chapter is to present a simple connection between the spectrum of the 'known term' $f(\cdot)$ (which is assumed to be almost-periodic) and the spectrum of the almost-periodic 'mild' solution of the differential equation $u' - Au = f$, where A is a generator of a C_0-semigroup of operators.

The result appears essentially, as a somewhat special case, in [10].

1. One considers a C_0-semigroup of operators $T(t)$ in the Banach space X, with infinitesimal generator A. An almost-periodic function $f(\cdot)$, $\mathbb{R} \to X$ is given, and then an almost-periodic function $u(\cdot)$, $\mathbb{R} \to X$, which is a 'mild' solution of the differential equation over $\mathbb{R}$

$$u'(t) - Au(t) = f(t)$$

that is it admits the representation formula

$$u(t) = T(t-a)u(a) + \int_a^t T(t-s)f(s)\,\mathrm{d}s\,, \qquad \forall a \in \mathbb{R},\ \forall t \geq a\,. \tag{1.1}$$

It is well known (see for instance [18]-p. 85) that, $\forall \lambda \in \mathbb{R}$, the mean values

$$a(\lambda,u) = \lim_{r\to\infty} \frac{1}{r}\int_0^r \mathrm{e}^{-\mathrm{i}\lambda t} u(t)\,\mathrm{d}t\,, \qquad a(\lambda,f) = \lim_{r\to\infty} \frac{1}{r}\int_0^r \mathrm{e}^{-\mathrm{i}\lambda t} f(t)\,\mathrm{d}t\,,$$

both exist. There is a connection between these mean values, which can be expressed in the following:

Theorem *For any $\lambda \in \mathbb{R}$, the mean value $a(\lambda,u)$ belongs to the domain $\mathcal{D}(A)$ and the equality*

$$(\mathrm{i}\lambda - A)a(\lambda,u) = a(\lambda,f) \tag{1.2}$$

holds true.

Remark For a similar result in a simpler situation see Example 1-p. 95 in [18].

The proof goes as follows. Let us make a change of variable in (1.1): $t-a=\sigma$, $t=a+\sigma$. Thus we get

$$u(a+\sigma) = T(\sigma)u(a) + \int_a^{a+\sigma} T(a+\sigma-s)f(s)\,\mathrm{d}s\,, \qquad \forall a \in \mathbb{R},\ \forall \sigma \geq 0,$$

whence one derives

$$T(\sigma)u(a) = u(a+\sigma) - \int_a^{a+\sigma} T(a+\sigma-s)f(s)\,\mathrm{d}s\,, \qquad \forall a \in \mathbb{R},\ \forall \sigma \geq 0. \tag{1.3}$$

Let us denote here by $\hat{u}(\lambda)$, $\hat{f}(\lambda)$ the mean values $a(\lambda,u)$, $a(\lambda,f)$. Then we get

$$T(\sigma)\hat{u}(\lambda) = T(\sigma)\lim_{r\to\infty}\frac{1}{r}\int_0^r \mathrm{e}^{-\mathrm{i}\lambda a}u(a)\,\mathrm{d}a = \lim_{r\to\infty}\frac{1}{r}\int_0^r \mathrm{e}^{-\mathrm{i}\lambda a}T(\sigma)u(a)\,\mathrm{d}a$$

(note that the function $a \in \mathbb{R} \to T(\sigma)u(a)$ is also almost-periodic, $\forall \sigma \geq 0$).

We next use (1.3) to obtain

$$T(\sigma)\hat{u}(\lambda) = \lim_{r\to\infty}\frac{1}{r}\int_0^r \mathrm{e}^{-\mathrm{i}\lambda a}\left[u(a+\sigma) - \int_a^{a+\sigma} T(a+\sigma-s)f(s)\,\mathrm{d}s\right]\mathrm{d}a$$

$$= \lim_{r\to\infty}\frac{1}{r}\int_0^r \mathrm{e}^{-\mathrm{i}\lambda a}u(a+\sigma)\,\mathrm{d}a - \lim_{r\to\infty}\frac{1}{r}\int_0^r\left(\int_a^{a+\sigma} T(a+\sigma-s)f(s)\,\mathrm{d}s\right)\cdot \mathrm{e}^{-\mathrm{i}\lambda a}\,da\,. \tag{1.4}$$

Then, note that

$$\int_0^r e^{-i\lambda a}u(a+\sigma)\,da = \left(\int_0^r e^{-i\lambda(a+\sigma)}u(a+\sigma)\,da\right)e^{i\lambda\sigma} = \left(\int_\sigma^{\sigma+r} e^{-i\lambda\xi}u(\xi)\,d\xi\right)e^{i\lambda\sigma}$$

and therefore

$$\lim_{r\to\infty}\frac{1}{r}\int_0^r e^{-i\lambda a}u(a+\sigma)\,da = e^{i\lambda\sigma}\lim_{r\to\infty}\frac{1}{r}\int_\sigma^{\sigma+r} e^{-i\lambda\xi}u(\xi)\,d\xi = e^{i\lambda\sigma}\hat{u}(\lambda)$$

(we used here (iv)-p. 87 of [18]).

Now we consider the second term in the right-hand side of (1.4); we see that

$$\int_0^r\left(\int_a^{a+\sigma} T(a+\sigma-s)f(s)\,ds\right)e^{-i\lambda a}\,da = (s = a+\tau \text{ in the internal integral})$$

$$\int_0^r\left(\int_0^\sigma T(\sigma-\tau)f(a+\tau)\,d\tau\right)e^{-i\lambda a}\,da$$

$$= \int_0^r\left(\int_0^\sigma T(\sigma-\tau)e^{i\lambda\tau}e^{-i\lambda(a+\tau)}f(a+\tau)\,d\tau\right)da$$

$$= \int_0^\sigma e^{i\lambda\tau}T(\sigma-\tau)\left(\int_0^r e^{-i\lambda(a+\tau)}f(a+\tau)\,da\right)d\tau$$

$$= \int_0^\sigma e^{i\lambda\tau}T(\sigma-\tau)\left(\int_\tau^{r+\tau} e^{-i\lambda u}f(u)\,du\right)d\tau\ ;$$

therefore

$$\frac{1}{r}\int_0^r\left(\int_a^{a+\sigma} T(a+\sigma-s)f(s)\,ds\right)e^{-i\lambda a}\,da$$

$$= \int_0^\sigma e^{i\lambda\tau}T(\sigma-\tau)\left(\frac{1}{r}\int_\tau^{r+\tau} e^{-i\lambda u}f(u)\,du\right)d\tau\ .$$

Next, note again that $\frac{1}{r}\int_{\tau}^{r+\tau} e^{-i\lambda u} f(u)\,du \to \hat{f}(\lambda)$ as $r \to \infty$, uniformly on $\tau \in \mathbb{R}$. It follows readily that

$$\lim_{r\to\infty} \frac{1}{r}\int_0^r \left(\int_a^{a+\sigma} T(a+\sigma-s) f(s)\,ds \right) e^{-i\lambda a}\,da = \int_0^\sigma e^{i\lambda\tau} T(\sigma-\tau)\hat{f}(\lambda)\,d\tau \,.$$

Previous computations give that

$$T(\sigma)\hat{u}(\lambda) = e^{i\lambda\sigma}\hat{u}(\lambda) - \int_0^\sigma e^{i\lambda\tau} T(\sigma-\tau)\hat{f}(\lambda)\,d\tau$$

hence

$$\frac{T(\sigma)-I}{\sigma}\hat{u}(\lambda) = \frac{e^{i\lambda\sigma}-1}{\sigma}\hat{u}(\lambda) - e^{i\lambda\sigma}\frac{1}{\sigma}\int_0^\sigma e^{-i\lambda(\sigma-\tau)} T(\sigma-\tau)\hat{f}(\lambda)\,d\tau \,.$$

In the right-hand side we first see that

$$\lim_{\sigma\to 0+} \frac{e^{i\lambda\sigma}-1}{\sigma} = i\lambda \,.$$

Next

$$\int_0^\sigma e^{-i\lambda(\sigma-\tau)} T(\sigma-\tau)\hat{f}(\lambda)\,d\tau = (\sigma-\tau=u) \int_0^\sigma e^{-i\lambda u} T(u)\hat{f}(\lambda)\,du \,,$$

so that

$$\frac{1}{\sigma}\int_0^\sigma e^{-i\lambda u} T(u)\hat{f}(\lambda)\,du \to \hat{f}(\lambda) \quad \text{as} \quad \sigma \to 0+ \,.$$

All this implies that $\lim_{\sigma\to 0+} \frac{T(\sigma)-I}{\sigma}\hat{u}(\lambda)$ exists and equals $i\lambda\hat{u}(\lambda) - \hat{f}(\lambda)$; hence $\hat{u}(\lambda) \in \mathcal{D}(A)$ and

$$A\hat{u}(\lambda) = i\lambda\hat{u}(\lambda) - \hat{f}(\lambda) \,, \quad \forall \lambda \in \mathbb{R}$$

which is (1.2).

Chapter XVI

Variational form of the equation $Au=f$ in a Hilbert space

Introduction

We consider here the equation $Au = f$ in a Hilbert space where A is a symmetric negative operator, of domain $\mathcal{D}(A) \subset H$. We establish a connection between its solution and the minimizer of the functional F, given by $F(v) = (-Av, v) - 2(v, f)$, $\forall v \in \mathcal{D}(A)$.

1. We start this discussion with the particular (but fundamental) example of the so-called 'Poisson equation': $\Delta u = f$, where $\Delta = \sum_1^n \frac{\partial^2}{\partial x_i^2}$. For instance, let us remember that in 12-Chapter II we considered the Hilbert space $L^2(\Omega)$, where Ω is a bounded 'regular' open set in $\mathbb{R}^n$, with boundary $S = \partial\Omega$. Let $\mathcal{D} = \{\varphi \in C^2(\overline{\Omega});\ \varphi(x) = 0,\ \forall x \in S\}$, and let the operator A, $\mathcal{D} \to L^2(\Omega)$, be given by $A\varphi = \Delta\varphi$, $\forall \varphi \in \mathcal{D}$. (Note that $A\varphi \in C^0(\overline{\Omega})$ which is contained in $L^2(\Omega)$.) We saw that $(A\varphi, \psi) = (\varphi, A\psi)$, $\forall \varphi, \psi \in \mathcal{D}$, that is A is a symmetric operator.

If we compute the $L^2(\Omega)$ scalar product (Au, u), for $u \in \mathcal{D}$, we find the following:

$$(Au,u) = \int_\Omega (\Delta u)(x) \cdot u(x)\,\mathrm{d}x = \int_\Omega \sum_{i=1}^n \frac{\partial}{\partial x_i}\left(\frac{\partial u}{\partial x_i} u\right) \mathrm{d}x$$

$$-\int_\Omega \left\{ \sum_{i=1}^n \left(\frac{\partial u}{\partial x_i}\right)^2 \right\} \mathrm{d}x = \int_S \left\{ u(x) \sum_{i=1}^n \frac{\partial u}{\partial x_i} v_i(x) \right\} \mathrm{d}\sigma$$

$$-\int_\Omega \left\{ \sum_{i=1}^n \left(\frac{\partial u}{\partial x_i}\right)^2 \right\} \mathrm{d}x = -\int_\Omega \left\{ \sum_{i=1}^n \left(\frac{\partial u}{\partial x_i}\right)^2 \right\} \mathrm{d}x$$

($v_i(x)$ are the components of the unit exterior normal $\vec{v}(x)$ to S at the point x; the surface integral vanishes since $u|_S = 0$).

Thus, we have found that $(Au, u) \le 0$, $\forall u \in \mathcal{D}$.

Let us assume also that Ω is a connected set in $\mathbb{R}^n$; if $(Au,u)=0$ for some $u \in \mathcal{D}$, it follows that each partial derivative $\frac{\partial u}{\partial x_i}$ vanishes in Ω; thus u =const in Ω, and $u=0$ in Ω. This means that $(Au,u)<0$, $\forall u \in \mathcal{D}$, $u \neq \mathbf{0}$; we say that A is a negative operator; accordingly, the operator $-A$ will be called a positive operator.

Next, for such an operator A we consider the equation $Au=f$, where f is given in $L^2(\Omega)$. We note the *uniqueness* of the solution:

if $Au_1 = f$ *and* $Au_2 = f$, *with* $u_1, u_2 \in \mathcal{D}$, *then* $u_1 = u_2$.

In fact, we see immediately that $A(u_1 - u_2) = \mathbf{0}$, whence $0 = \left(A(u_1 - u_2),\ u_1 - u_2\right)$. This entails, as seen above, that $u_1 - u_2 = \mathbf{0}$, $u_1 = u_2$.

We next prove:

Theorem 1 *Let us assume that* $u \in \mathcal{D}(A)$ *is a solution of the equation* $-Au = f$ *where f is given in H. Then the functional F;* $\mathcal{D}(A) \to \mathbb{R}$, *given by the formula*

$$F(v) = (-Av, v) - 2(v, f) \tag{1.1}$$

takes its minimum value for $v = u$: $F(v) \geq F(u)$, $\forall v \in \mathcal{D}(A)$.

Remark In the 'concrete' case of Poisson's equation $\Delta u = f$ which has been discussed above we have obviously, that

$$F(v) = \int_\Omega \left\{ \sum_1^n \left(\frac{\partial v}{\partial x_i} \right)^2 \right\} \mathrm{d}x - 2\int_\Omega v(x) f(x)\, \mathrm{d}x \ .$$

Proof Let us assume (1.1) and also $-Au = f$. It follows that

$$F(v) = (-Av, v) + 2(v, Au) = (-Av, v) + (Av, u) + (v, Au)$$

(using the symmetry of operator A). Therefore

$$F(v) = (-A(v-u),\ v-u) + (Au, u)$$

(remember that H is a Hilbert space over $\mathbb{R}$, so that

$(Au,v)=(v,Au)$)

and accordingly we get:

$$F(v) \geq (Au,u) \ , \qquad \forall v \in \mathcal{D}(A) \ ;$$

however $(Au,u) = F(u)$, as is easily seen, so the theorem is proved. ■

The following converse result is also of interest.

Theorem 2 *Let us assume that the element u_0 of $\mathcal{D}(A)$ - which is dense in H^- minimizes the functional F on $\mathcal{D}(A)$. Then $-Au_0 = f$.*

Proof Let us consider the function ϕ, $\mathbb{R} \to \mathbb{R}$, given by $\phi(t) = F(u_0 + tw)$, where w is fixed in $\mathcal{D}(A)$. We know by assumption that

$$\phi(t) = F(u_0 + tw) \geq F(u_0) = \phi(0) \ , \qquad \forall t \in \mathbb{R} \ . \tag{1.2}$$

On the other hand, we may note also that $\phi(\cdot)$ is a second-degree polynomial in the variable t:

$$\begin{aligned}\phi(t) &= (-A(u_0 + tw),\ u_0 + tw) - 2(u_0 + tw,\ f) \\ &= (-Au_0, u_0) + t[(-Aw, u_0) - (Au_0, w) - 2(w, f)] + t^2[(-Aw, w)] - 2(u_0, f) \ .\end{aligned}$$

Therefore, $t = 0$ is a minimum point for ϕ and therefore $\phi'(0) - 0$. This means that

$$(-Aw, u_0) - (Au_0, w) - 2(w, f) = 0 \ , \qquad (Au_0, w) + (f, w) = 0 \ ,$$
$$(Au_0 + f, w) = 0 \ , \qquad \forall w \in \mathcal{D}(A) \ .$$

As $\mathcal{D}(A)$ is here dense in H, we obtain $Au_0 = -f$.

Remark In the special case explained at the beginning, $\mathcal{D}(A) = \mathcal{D} = \{\varphi \in C^2(\overline{\Omega}), \varphi|_S = 0\}$.

This set is known to be dense in $L^2(\Omega)$, so Theorem 2 applies and a minimizer $u_0 \in \mathcal{D}$ of the functional F will be a solution of Poisson's equation $-\Delta u = f$.

References

[1] Brezis, H. (1983), *Analyse fonctionnelle (Théorie et applications),* Paris, New York, Milan: Masson.

[2] Dunford, N. and Schwartz, J.T. (1958), *Linear Operators, Part I: General Theory*, New York: Interscience Publishers, Inc.

[3] Friedman, A. (1982), *Foundations of Modern Analysis*, New York: Dover Publications, Inc.

[4] Goldberg, S. (1966), *Unbounded Linear Operators*, New York, Toronto, London: McGraw-Hill.

[5] Hille, E. and Phillips, R.S. (1957), *Functional Analysis and Semigroups*, Providence, RI: American Mathematical Society.

[6] Hirsch, M.W. and Smale, S. (1974), *Differential Equations, Dynamical Systems and Linear Algebra*, New York, San Francisco, London: Academic Press.

[7] Levine, H.A. (1975), Uniqueness and growth of weak solutions to certain linear differential equations in Hilbert space, *Journal of Differential Equations,* **17**(1), 73-81.

[8] Leviton, B.M. and Zhikov, V.V. (1982), *Almost Periodic Functions and Differential Equations*, Cambridge, London, New York: Cambridge University Press.

[9] Petersen, B.E. (1983), *Introduction to the Fourier Transform and Pseudo-Differential Operators*, Boston, London, Melbourne: Pitman Publishing.

[10] Priiss, J. (1984), On the spectrum of C_0-semigroups, *Trans. Amer. Math. Soc.,* **284**, 847-857.

[11] Reed, M. and Simon, B. (1972), *Methods of Modern Mathematical Physics I*, New York, London: Academic Press.

[12] Walker, J.A. (1980), *Dynamical Systems and Evolution Equations*, New York, London: Plenum Press.

[13] Yosida, K. (1965), *Functional Analysis*, Berlin: Springer-Verlag.

[14] Young, N. (1988), *An Introduction to Hilbert Space*, Cambridge, New York, Sydney: Cambridge University Press.

[15] Zaidman, S. (1969), Uniqueness of bounded weak solutions for some abstract differential equations, *Annali dell'Università di Ferrara,* **XIV**(9), 97-9.

[16] Zaidman, S. (1976), An asymptotic result for weak differential inequalities, *Rend. Semin. Met. Univ. Padova,* **55**, 167-178.

[17] Zaidman, S. (1979), *Abstract Differential Equations*, San Francisco, London, Melbourne: Pitman Publishing, RNM 36.

[18] Zaidman, S. (1985), *Almost-Periodic Functions in Abstract Spaces*, Boston, London, Melbourne, Pitman Publishing, RNM 126.

[19] Zaidman, S. (1989), Some remarks on the abstract Cauchy problem in the ultraweak sense. *Proceedings of KIT Mathematics Workshop,* **4**, 127-143.

[20] Zaidman, S. (1989), *Une introduction à le théorie des équations aux dérivés partielles*, CRM-Université de Montréal.

[21] Zaidman, S. (1994) (1995), *Topics in Abstract Differential Equations I,II*, Longman Scientific and Technical, Pitman Publishing, RNM 304, 321.

[22] Zaidman, S. (1997), *Advanced Calculus (An Introduction to Mathematical Analysis)*, Singapore, New Jersey, London: World Scientific.

Subject Index

M

N

O

S

T

U

V

W

Z

Index of Symbols

V	linear or vector space
$\mathbb{R}$	the field of real numbers
$\mathbb{C}$	the field of complex numbers
(x,y)	scalar or inner product
K	real or complex field
$\mathbf{0}$	the zero-element in V
$\mathbb{C}^n$	Cartesian product of $\mathbb{C}$ n times
$C[a,b]$	the linear space of all continuous complex-valued functions on $[a,b]$
$x \perp y$	x is orthogonal to y, $(x,y)=0$
$\|x\|$	norm of element x
$d(x,y)$	distance in metric space
$\mathfrak{M}^{\perp}$	orthogonal complement
$B(x_0,\rho)$	open ball in normed space
$C(x_0,\rho)$	closed ball in normed space
E'	dual space to E
$\mathcal{L}(E,K)$	space of linear continuous mappings from E into K
Ker F	kernel space of F
$B(x,y)$	sesquilinear form on Cartesian product of Hilbert spaces
$\mathcal{L}(E,F)$	linear continuous operators from E into F
$T(X)$	image of space X under operator T
$\{T_\alpha\}_{\alpha\in\mathcal{A}}$	family of elements in $\mathcal{L}(X,Y)$, indexed by the set $\mathcal{A}$
U.B.T.	uniform boundedness theorem

$(T_n)_1^\infty$	sequence of elements in $\mathcal{L}(X,Y)$
$\mathcal{L}(H)$	same as $\mathcal{L}(H,H)$
$\{F_y\}_{\|y\|\le 1}$	family of linear functionals on H, indexed by the closed unit ball
$A(t)$	function from $[a,b]$ to $\mathcal{L}(X)$
A^*	Hilbert adjoint of operator A
A^{**}	same as $(A^*)^*$ - double adjoint of A
$h \perp \mathcal{H}_1$	$(h,k)=0 \quad \forall k \in \mathcal{H}_1$
$a(u,v)$	sesquilinear form on $\mathcal{H} \times \mathcal{H}$
$a^*(u,v)$	adjoint sesquilinear form
$\mathcal{D}^*$	domain of the Hilbert adjoint to the unbounded operator A
$L_1 \subset L_2$	inclusion of linear operators
$L^2(\mathbb{R})$	space of square integrable complex-valued functions on $\mathbb{R}$
$L^1(\mathbb{R})$	space of absolutely Lebesgue integrable functions on $\mathbb{R}$
$C_0(\mathbb{R})$	complex-valued continuous functions with compact support on $\mathbb{R}$
a.e	almost-everywhere
$G(A)$	graph of linear operator A
$R(A)$	range of linear operator A: $\{Au\}_{u\in\mathcal{D}(A)}$
Ker A	set $\{u \in \mathcal{D}(A),\ Au = \mathbf{0}\}$
$\tilde{T}$	closed extension of operator T
$\overline{T}$	closure of operator T
$L^2(\Omega)$	space of square-integrable complex-valued functions on the open subset Ω of $\mathbb{R}^n$
$C^1(\Omega)$	continuously differentiable functions on Ω
$C_0^1(\Omega)$	continuously differentiable functions with compact support in Ω
$(G_T)^{\perp\perp}$	same as $\left((G_T)^\perp\right)^\perp$
T^{***}	triple adjoint, same as $(T^{**})^*$
$\partial\Omega$	boundary of the open set Ω in $\mathbb{R}^n$

$C^2(\overline{\Omega})$	twice continuously differentiable functions on $\overline{\Omega}$
Δ	Laplace operator $= \sum_{i=1}^{n} \frac{\partial^2}{\partial x_i^2}$
$\rho(T)$	resolvent set of operator T
$R_\lambda(T) = R(\lambda, T)$	resolvent operator of T: $(\lambda I - T)^{-1}$
$\sigma(T)$	spectrum set of operator T
$\sigma_r(T)$	residual spectrum of T
$\sigma_p(T)$	point spectrum of T
$\sigma_c(T)$	continuous spectrum of T
l^2	space of real-valued square summable sequences
A'	dual operator to A
X''	bidual space to X: $(X')' = X''$
$\mathcal{J}$	canonical mapping, X into X''
X'''	third dual space $= (X'')'$
x', x'', x'''	elements of dual spaces X', X'', X'''
$u(t), u'(t)$	function from a real interval into a B-space; its derivative
$x'_+(t_0)$	right-strong derivative of function $x(t)$, for $t = t_0$
$x'(t_0)$	strong derivative of function $x(\cdot)$, for $t = t_0$
$C(I; X)$	space of continuous functions $x(\cdot)$, from the real interval I into the B-space X
$U(t)$	family of operators associated to solutions of equation $Bu'(t) = Au(t)$, $u(0) = u_0$ on $[0, \infty)$
$\tilde{U}(t)$	extension of $U(t)$ to the whole space X
$C^1(I; X)$	space of functions with a continuous derivative, $I \to X$
C_0	strongly continuous semigroups of operators on $[0, \infty)$
$\exp(At)$	abstract exponential function
$\int_a^b f = \int_a^b f(\sigma)\, d\sigma$	Riemann integral of function f, $[a, b] \to X$
P_M	projection operator on subspace M

$M \perp N$	$(m,n)=0 \quad \forall m \in M,\ \forall n \in \mathbb{N}$
$N \ominus M$	orthogonal difference: $x \in N,\ x \perp M$
$A \geq B$	order relation between symmetric operators
$A^{1/2}$	square root of operator A
$\mathcal{X}$	metric space
x_{e}	equilibrium point
$\{S(t)\}_{t\geq 0}$	strongly continuous semigroup of $\mathcal{X}$
$H,\mathcal{H}$	Hilbert spaces
$\mathcal{M}(x,y)$	symmetric bilinear form on $D \times D$
$E(t)$	energy
X	Banach space
$\mathcal{D}(A)$	domain of operator A
$K_{A'}(0,\infty)$	test-functions
$\{\alpha_n(\cdot)\}$	sequence of mollifiers
$\dot{\alpha}_n(\cdot)$	derivative - alternative notation
$\int_a^\infty x(t)\,\mathrm{d}t$	improper Riemann integral for X-valued function
$\int_0^\infty \mathrm{e}^{-\lambda t} u(t)\,\mathrm{d}t$	Laplace transform of X-valued function $u(\cdot)$
$K_{A^*}(\mathbb{R})$	vector-valued test-functions on $\mathbb{R}$
$G(t)$	strongly continuous group of operators
$T(t)$	C_0-semigroup of operators
$K_{A^*}(0,\infty)$	test-functions
$\hat{\xi}(\tau)$	$=\frac{1}{\sqrt{2\pi}}\int_{\mathbb{R}} \mathrm{e}^{-\mathrm{i}\tau t}\xi(t)\,\mathrm{d}t =$ Fourier transform of $\xi(\cdot)$
$a(\lambda,u)=\hat{u}(\lambda)$	mean-value of function $u(\cdot)$
$C^0(\overline{\Omega})$	continuous functions on $\overline{\Omega}$
$\varphi\vert_S = 0$	functions $\overline{\Omega} \to \mathbb{C}$ such that they vanish on $S=\partial\Omega$